Pro Hadoop Data Analytics

Designing and Building Big Data Systems
using the Hadoop Ecosystem

Kerry Koitzsch

Apress®

Pro Hadoop Data Analytics: Designing and Building Big Data Systems using the Hadoop Ecosystem

Kerry Koitzsch
Sunnyvale, California, USA

ISBN-13 (pbk): 978-1-4842-1909-6   ISBN-13 (electronic): 978-1-4842-1910-2
DOI 10.1007/978-1-4842-1910-2

Library of Congress Control Number: 2016963203

Managing Director: Welmoed Spahr
Lead Editor: Celestin Suresh John
Technical Reviewer: Simin Boschma
Editorial Board: Steve Anglin, Pramila Balan, Laura Berendson, Aaron Black, Louise Corrigan, Jonathan Gennick, Robert Hutchinson, Celestin Suresh John, Nikhil Karkal, James Markham, Susan McDermott, Matthew Moodie, Natalie Pao, Gwenan Spearing
Coordinating Editor: Prachi Mehta
Copy Editor: Larissa Shmailo
Compositor: SPi Global
Indexer: SPi Global
Artist: SPi Global

Distributed to the book trade worldwide by Springer Science+Business Media New York, 233 Spring Street, 6th Floor, New York, NY 10013. Phone 1-800-SPRINGER, fax (201) 348-4505, e-mail orders-ny@springer-sbm.com, or visit www.springeronline.com. Apress Media, LLC is a California LLC and the sole member (owner) is Springer Science + Business Media Finance Inc (SSBM Finance Inc). SSBM Finance Inc is a **Delaware** corporation.

For information on translations, please e-mail rights@apress.com, or visit www.apress.com.

Apress and friends of ED books may be purchased in bulk for academic, corporate, or promotional use. eBook versions and licenses are also available for most titles. For more information, reference our Special Bulk Sales–eBook Licensing web page at www.apress.com/bulk-sales.

Any source code or other supplementary materials referenced by the author in this text are available to readers at www.apress.com. For detailed information about how to locate your book's source code, go to www.apress.com/source-code/. Readers can also access source code at SpringerLink in the Supplementary Material section for each chapter.

Printed on acid-free paper

To Sarvnaz, whom I love.

Contents at a Glance

Contents

About the Author

Kerry Koitzsch has had more than twenty years of experience in the computer science, image processing, and software engineering fields, and has worked extensively with Apache Hadoop and Apache Spark technologies in particular. Kerry specializes in software consulting involving customized big data applications including distributed search, image analysis, stereo vision, and intelligent image retrieval systems. Kerry currently works for Kildane Software Technologies, Inc., a robotic systems and image analysis software provider in Sunnyvale, California.

About the Technical Reviewer

Simin Boschma has over twenty years of experience in computer design engineering. Simin's experience also includes program and partner management, as well as developing commercial hardware and software products at high-tech companies throughout Silicon Valley, including Hewlett-Packard and SanDisk. In addition, Simin has more than ten years of experience in technical writing, reviewing, and publication technologies. Simin currently works for Kildane Software Technologies, Inc. in Sunnyvale, CA.

Acknowledgments

I would like to acknowledge the invaluable help of my editors Celestin Suresh John and Prachi Mehta, without whom this book would never have been written, as well as the expert assistance of the technical reviewer Simin Bochma.

Introduction

The Apache Hadoop software library has come into it's own. It is the basis for advanced distributed development for a host of companies, government institutions, and scientific research facilities. The Hadoop ecosystem now contains dozens of components for everything from search, databases, and data warehousing to image processing, deep learning, and natural language processing. With the advent of Hadoop 2, different resource managers may be used to provide an even greater level of sophistication and control than previously possible. Competitors, replacements, as well as successors and mutations of the Hadoop technologies and architectures abound. These include Apache Flink, Apache Spark, and many others. The "death of Hadoop" has been announced many times by software experts and commentators.

We have to face the question squarely: is Hadoop dead? It depends on the perceived boundaries of Hadoop itself. Do we consider Apache Spark, the in-memory successor to Hadoop's batch file approach, a part of the Hadoop family simply because it also uses HDFS, the Hadoop file system? Many other examples of "gray areas" exist in which newer technologies replace or enhance the original "Hadoop classic" features. Distributed computing is a moving target and the boundaries of Hadoop and its ecosystem have changed remarkably over a few short years. In this book, we attempt to show some of the diverse and dynamic aspects of Hadoop and its associated ecosystem, and to try to convince you that, although changing, Hadoop is still very much alive, relevant to current software development, and particularly interesting to data analytics programmers.

PART I

Concepts

The first part of our book describes the basic concepts, structure, and use of the distributed analytics software system, why it is useful, and some of the necessary tools required to use this type of distributed system. We will also introduce some of the distributed infrastructure we need to build systems, including Apache Hadoop and its ecosystem.

CHAPTER 1

■ ■ ■

Overview: Building Data Analytic Systems with Hadoop

This book is about designing and implementing software systems that ingest, analyze, and visualize big data sets. Throughout the book, we'll use the acronym BDA or BDAs (big data analytics system) to describe this kind of software. Big data itself deserves a word of explanation. As computer programmers and architects, we know that what we now call "big data" has been with us for a very long time—decades, in fact, because "big data" has always been a relative, multi-dimensional term, a space which is not defined by the mere size of the data alone. Complexity, speed, veracity—and of course, size and volume of data—are all dimensions of any modern "big data set".

In this chapter, we discuss what big data analytic systems (BDAs) using Hadoop are, why they are important, what data sources, sinks, and repositories may be used, and candidate applications which are—and are not—suitable for a distributed system approach using Hadoop. We also briefly discuss some alternatives to the Hadoop/Spark paradigm for building this type of system.

There has always been a sense of urgency in software development, and the development of big data analytics is no exception. Even in the earliest days of what was to become a burgeoning new industry, big data analytics have demanded the ability to process and analyze more and more data at a faster rate, and at a deeper level of understanding. When we examine the practical nuts-and-bolts details of software system architecting and development, the fundamental requirement to process more and more data in a more comprehensive way has always been a key objective in abstract computer science and applied computer technology alike. Again, big data applications and systems are no exception to this rule. This can be no surprise when we consider how available global data resources have grown explosively over the last few years, as shown in Figure 1-1.

© Kerry Koitzsch 2017

K. Koitzsch, *Pro Hadoop Data Analytics*, DOI 10.1007/978-1-4842-1910-2_1

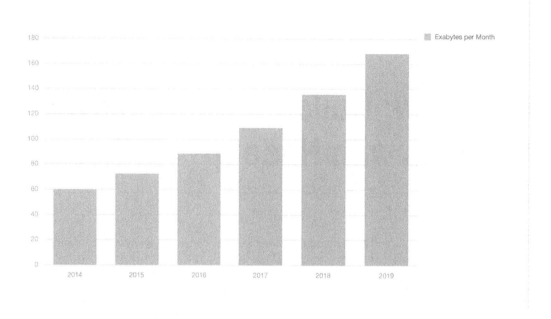

Figure 1-1. *Annual data volume statistics [Cisco VNI Global IP Traffic Forecast 2014-2019]*

As a result of the rapid evolution of software components and inexpensive off-the-shelf processing power, combined with the rapidly increasing pace of software development itself, architects and programmers desiring to build a BDA for their own application can often feel overwhelmed by the technological and strategic choices confronting them in the BDA arena. In this introductory chapter, we will take a high-level overview of the BDA landscape and attempt to pin down some of the technological questions we need to ask ourselves when building BDAs.

1.1   A Need for Distributed Analytical Systems

We need distributed big data analysis because old-school business analytics are inadequate to the task of keeping up with the volume, complexity, variety, and high data processing rates demanded by modern analytical applications. The big data analysis situation has changed dramatically in another way besides software alone. Hardware costs—for computation and storage alike—have gone down tremendously. Tools like Hadoop, which rely on clusters of relatively low-cost machines and disks, make distributed processing a day-to-day reality, and, for large-scale data projects, a necessity. There is a lot of support software (frameworks, libraries, and toolkits) for doing distributed computing, as well. Indeed, the problem of choosing a technology stack has become a serious issue, and careful attention to application requirements and available resources is crucial.

Historically, hardware technologies defined the limits of what software components are capable of, particularly when it came to data analytics. Old-school data analytics meant doing statistical visualization (histograms, pie charts, and tabular reports) on simple file-based data sets or direct connections to a relational data store. The computational engine would typically be implemented using batch processing on a single server. In the brave new world of distributed computation, the use of a cluster of computers to divide and conquer a big data problem has become a standard way of doing computation: this scalability allows us to transcend the boundaries of a single computer's capabilities and add as much off-the-shelf hardware as we need (or as we can afford). Software tools such as Ambari, Zookeeper, or Curator assist us in managing the cluster and providing scalability as well as high availability of clustered resources.

1.2 The Hadoop Core and a Small Amount of History

Some software ideas have been around for so long now that it's not even computer history any more—it's computer archaeology. The idea of the "map-reduce" problem-solving method goes back to the second-oldest programming language, LISP (List Processing) dating back to the 1950s. "Map," "reduce." "send," and "lambda" were standard functions within the LISP language itself! A few decades later, what we now know as Apache Hadoop, the Java-based open source-distributed processing framework, was not set in motion "from scratch." It evolved from Apache Nutch, an open source web search engine, which in turn was based on Apache Lucene. Interestingly, the R statistical library (which we will also be discussing in depth in a later chapter) also has LISP as a fundamental influence, and was originally written in the LISP language.

The Hadoop Core component deserves a brief mention before we talk about the Hadoop ecosystem. As the name suggests, the Hadoop Core is the essence of the Hadoop framework [figure 1.1]. Support components, architectures, and of course the ancillary libraries, problem-solving components, and sub-frameworks known as the Hadoop ecosystem are all built on top of the Hadoop Core foundation, as shown in Figure 1-2. Please note that within the scope of this book, we will not be discussing Hadoop 1, as it has been supplanted by the new reimplementation using YARN (Yet Another Resource Negotiator). Please note that, in the Hadoop 2 system, MapReduce has not gone away, it has simply been modularized and abstracted out into a component which will play well with other data-processing modules.

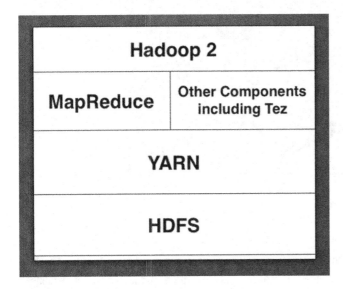

Figure 1-2. *Hadoop 2 Core diagram*

1.3 A Survey of the Hadoop Ecosystem

Hadoop and its ecosystem, plus the new frameworks and libraries which have grown up around them, continue to be a force to be reckoned with in the world of big data analytics. The remainder of this book will assist you in formulating a focused response to your big data analytical challenges, while providing a minimum of background and context to help you learn new approaches to big data analytical problem solving. Hadoop and its ecosystem are usually divided into four main categories or functional blocks as shown in Figure 1-3. You'll notice that we include a couple of extra blocks to show the need for software "glue" components as well as some kind of security functionality. You may also add support libraries and frameworks to your BDA system as your individual requirements dictate.

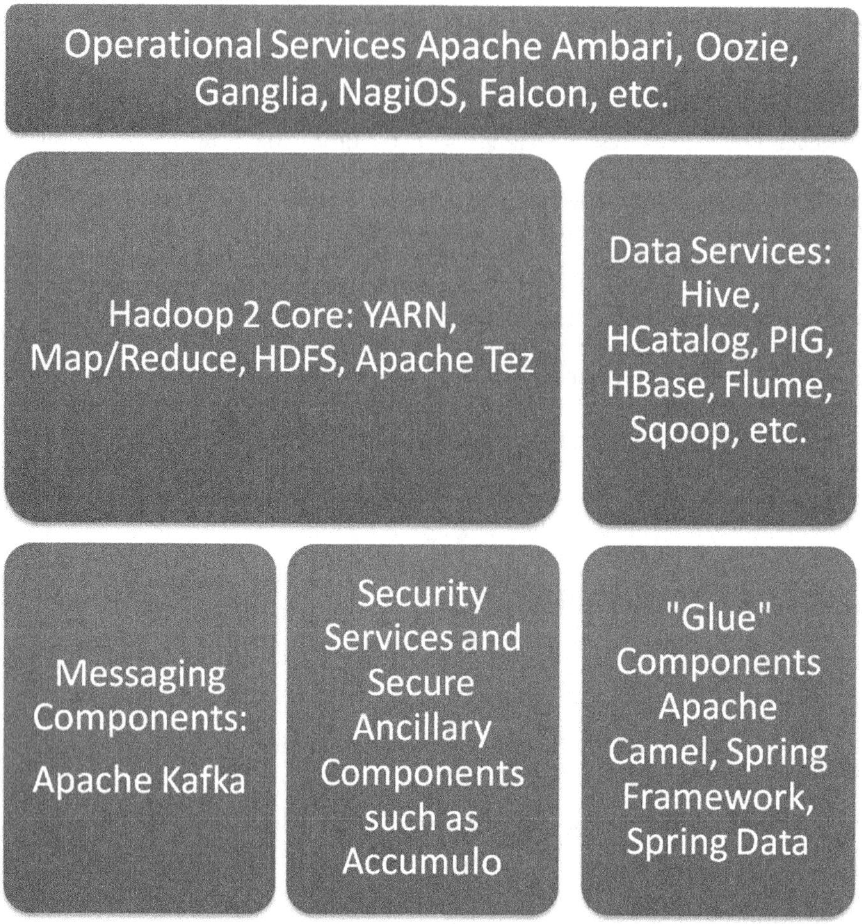

Operational Services Apache Ambari, Oozie, Ganglia, NagiOS, Falcon, etc.	

Hadoop 2 Core: YARN, Map/Reduce, HDFS, Apache Tez	Data Services: Hive, HCatalog, PIG, HBase, Flume, Sqoop, etc.

Messaging Components: Apache Kafka	Security Services and Secure Ancillary Components such as Accumulo	"Glue" Components Apache Camel, Spring Framework, Spring Data

Figure 1-3. *Hadoop 2 Technology Stack diagram*

▪ **Note** Throughout this book we will keep the emphasis on free, third-party components such as the Apache components and libraries mentioned earlier. This doesn't mean you can't integrate your favorite graph database (or relational database, for that matter) as a data source into your BDAs. We will also emphasize the flexibility and modularity of the open source components, which allow you to hook data pipeline components together with a minimum of additional software "glue." In our discussion we will use the Spring Data component of the Spring Framework, as well as Apache Camel, to provide the integrating "glue" support to link our components.

1.4 AI Technologies, Cognitive Computing, Deep Learning, and Big Data Analysis

Big data analysis is not just simple statistical analysis anymore. As BDAs and their support frameworks have evolved, technologies from machine learning (ML) artificial intelligence (AI), image and signal processing, and other sophisticated technologies (including the so-called "cognitive computing" technologies) have matured and become standard components of the data analyst's toolkit.

1.5 Natural Language Processing and BDAs

Natural language processing (NLP) components have proven to be useful in a large and varied number of domains, from scanning and interpreting receipts and invoices to sophisticated processing of prescription data in pharmacies and medical records in hospitals, as well as many other domains in which unstructured and semi-structured data abounds. Hadoop is a natural choice when processing this kind of "mix-and-match" data source, in which bar codes, signatures, images and signals, geospatial data (GPS locations) and other data types might be thrown into the mix. Hadoop is also a very powerful means of doing large-scale document analyses of all kinds.

We will discuss the so-called "semantic web" technologies, such as taxonomies and ontologies, rule-based control, and NLP components in a separate chapter. For now, suffice it to say that NLP has moved out of the research domain and into the sphere of practical app development, with a variety of toolkits and libraries to choose from. Some of the NLP toolkits we'll be discussing in this book are the Python-based Natural Language Toolkit (NLTK), Stanford NLP, and Digital Pebble's Behemoth, an open source platform for large-scale document analysis, based on Apache Hadoop.[1]

1.6 SQL and NoSQL Querying

Data is not useful unless it is queried. The process of querying a data set—whether it be a key-value pair collection, a relational database result set from Oracle or MySQL, or a representation of vertices and edges such as that found in a graph database like Neo4j or Apache Giraph—requires us to filter, sort, group, organize, compare, partition, and evaluate our data. This has led to the development of query languages such as SQL, as well as all the mutations and variations of query languages associated with "NoSQL" components and databases such as HBase, Cassandra, MongoDB, CouchBase, and many others. In this book, we will concentrate on using read-eval-print loops (REPLs), interactive shells (such as IPython) and other interactive tools to express our queries, and we will try to relate our queries to well-known SQL concepts as much as possible, regardless of what software component they are associated with. For example, some graph databases such as Neo4j (which we will discuss in detail in a later chapter) have their own SQL-like query languages. We will try and stick to the SQL-like query tradition as much as possible throughout the book, but will point out some interesting alternatives to the SQL paradigm as we go.

---

[1]One of the best introductions to the "semantic web" approach is Dean Allemang and Jim Hendler's "Semantic Web for the Working Ontologist: Effective Modeling in RDFS and OWL", 2008, Morgan-Kaufmann/Elsevier Publishing, Burlington, MA. ISBN 978-0-12-373556-0.

1.7   The Necessary Math

In this book, we will keep the mathematics to a minimum. Sometimes, though, a mathematical equation becomes more than a necessary evil. Sometimes the best way to understand your problem and implement your solution is the mathematical route—and, again, in some situations the "necessary math" becomes the key ingredient for solving the puzzle. Data models, neural nets, single or multiple classifiers, and Bayesian graph techniques demand at least some understanding of the underlying dynamics of these systems. And, for programmers and architects, the necessary math can almost always be converted into useful algorithms, and from there to useful implementations.

1.8   A Cyclic Process for Designing and Building BDA Systems

There is a lot of good news when it comes to building BDAs these days. The advent of Apache Spark with its in-memory model of computation is one of the major positives, but there are several other reasons why building BDAs has never been easier. Some of these reasons include:

- a wealth of frameworks and IDEs to aid with development;

- mature and well-tested components to assist building BDAs, and corporation-supported BDA products if you need them. Framework maturity (such as the Spring Framework, Spring Data subframework, Apache Camel, and many others) has helped distributed system development by providing reliable core infrastructure to build upon.

- a vital online and in-person BDA development community with innumerable developer forums and meet-ups. Chances are if you have encountered an architectural or technical problem in regard to BDA design and development, someone in the user community can offer you useful advice.

Throughout this book we will be using the following nine-step process to specify and create our BDA example systems. This process is only suggestive. You can use the process listed below as-is, make your own modifications to it, add or subtract structure or steps, or come up with your own development process. It's up to you. The following steps have been found to be especially useful for planning and organizing BDA projects and some of the questions that arise as we develop and build them.

You might notice that problem and requirement definition, implementation, testing, and documentation are merged into one overall process. The process described here is ideally suited for a rapid-iteration development process where the requirements and technologies used are relatively constant over a development cycle.

The basic steps when defining and building a BDA system are as follows. The overall cycle is shown in Figure 1.4.

Figure 1-4. *A cyclic process for designing and building BDAs*

1. **Identify requirements for the BDA system.** The initial phase of development
   requires generating an outline of the technologies, resources, techniques and
   strategies, and other components necessary to achieve the objectives. The initial
   set of objectives (subject to change of course) need to be pinned down, ordered,
   and well-defined. It's understood that the objectives and other requirements
   are subject to change as more is learned about the project's needs. BDA systems
   have special requirements (which might include what's in your Hadoop cluster,
   special data sources, user interface, reporting, and dashboard requirements).
   Make a list of data source types, data sink types, necessary parsing,
   transformation, validation, and data security concerns. Being able to adapt
   your requirements to the plastic and changeable nature of BDA technologies
   will insure you can modify your system in a modular, organized way. Identify
   computations and processes in the components, determine whether batch or
   stream processing (or both) is required, and draw a flowchart of the computation
   engine. This will help define and understand the "business logic" of the system.

2. **Define the initial technology stack.** The initial technology stack will include a Hadoop Core as well as appropriate ecosystem components appropriate for the requirements you defined in the last step. You may include Apache Spark if you require streaming support, or you're using one of the machine learning libraries based on Spark we discuss later in the book. Keep in mind the programming languages you will use. If you are using Hadoop, the Java language will be part of the stack. If you are using Apache Spark, the Scala language will also be used. Python has a number of very interesting special applications, as we will discuss in a later chapter. Other language bindings may be used if they are part of the requirements.

3. **Define data sources, input and output data formats, and data cleansing processes.** In the requirement-gathering phase (step 0), you made an initial list of the data source/sink types and made a top-level flowchart to help define your data pipeline. A lot of exotic data sources may be used in a BDA system, including images, geospatial locations, timestamps, log files, and many others, so keep a current list of data source (and data sink!) types handy as you do your initial design work.

4. **Define, gather, and organize initial data sets.** You may have initial data for your project, test and training data (more about training data later in the book), legacy data from previous systems, or no data at all. Think about the minimum amount of data sets (number, kind, and volume) and make a plan to procure or generate the data you need. Please note that as you add new code, new data sets may be required in order to perform adequate testing. The initial data sets should exercise each module of the data pipeline, assuring that end-to-end processing is performed properly.

5. **Define the computations to be performed.** Business logic in its conceptual form comes from the requirements phase, but what this logic is and how it is implemented will change over time. In this phase, define inputs, outputs, rules, and transformations to be performed on your data elements. These definitions get translated into implementation of the computation engine in step 6.

6. **Preprocess data sets for use by the computation engine.** Sometimes the data sets need preprocessing: validation, security checks, cleansing, conversion to a more appropriate format for processing, and several other steps. Have a checklist of preprocessing objectives to be met, and continue to pay attention to these issues throughout the development cycle, and make necessary modifications as the development progresses.

7. **Define the computation engine steps; define the result formats**. The business logic, flow, accuracy of results, algorithm and implementation correctness, and efficiency of the computation engine will always need to be questioned and improved.

8. **Place filtered results in results repositories of data sinks**. Data sinks are the data repositories that hold the final output of our data pipeline. There may be several steps of filtration or transformation before your output data is ready to be reported or displayed. The final results of your analysis can be stored in files, databases, temporary repositories, reports, or whatever the requirements dictate. Keep in mind user actions from the UI or dashboard may influence the format, volume, and presentation of the outputs. Some of these interactive results may need to be persisted back to a data store. Organize a list of requirements specifically for data output, reporting, presentation, and persistence.

9. **Define and build output reports, dashboards, and other output displays and controls.** The output displays and reports, which are generated, provide clarity on the results of all your analytic computations. This component of a BDA system is typically written, at least in part, in JavaScript and may use sophisticated data visualization libraries to assist different kinds of dashboards, reports, and other output displays.

10. **Document, test, refine, and repeat.** If necessary, we can go through the steps again after refining the requirements, stack, algorithms, data sets, and the rest. Documentation initially consists of the notes you made throughout the last seven steps, but needs to be refined and rewritten as the project progresses. Tests need to be created, refined, and improved throughout each cycle. Incidentally, each development cycle can be considered a version, iteration, or however you like to organize your program cycles.

There you have it. A systematic use of this iterative process will enable you to design and build BDA systems comparable to the ones described in this book.

1.9 How The Hadoop Ecosystem Implements Big Data Analysis

The Hadoop ecosystem implements big data analysis by linking together all the necessary ingredients for analysis (data sources, transformations, infrastructure, persistence, and visualization) in a data pipeline architecture while allowing these components to operate in a distributed way. Hadoop Core (or in certain cases Apache Spark or even hybrid systems using Hadoop and Storm together) supplies the distributed system infrastructure and cluster (node) coordination via components such as ZooKeeper, Curator, and Ambari. On top of Hadoop Core, the ecosystem provides sophisticated libraries for analysis, visualization, persistence, and reporting.

The Hadoop ecosystem is more than tacked-on libraries to the Hadoop Core functionality. The ecosystem provides integrated, seamless components with the Hadoop Core specifically designed for solving specific distributed problems. For example, Apache Mahout provides a toolkit of distributed machine learning algorithms.

Having some well-thought-out APIs makes it easy to link up our data sources to our Hadoop engine and other computational elements. With the "glue" capability of Apache Camel, Spring Framework, Spring Data, and Apache Tika, we will be able to link up all our components into a useful dataflow engine.

1.10 The Idea of "Images as Big Data" (IABD)

Images—pictures and signals of all kinds in fact—are among the most widespread, useful, and complex sources of "big data type" information.

Images are sometimes thought of as two-dimensional arrays of atomic units called pixels and, in fact (with some associated metadata), this is usually how images are represented in computer programming languages such as Java, and in associated image processing libraries such as Java Advanced Imaging (JAI), OpenCV and BoofCV, among others. However, biological systems "pull things out" of these "two-dimensional arrays": lines and shapes, color, metadata and context, edges, curves, and the relationships between all these. It soon becomes apparent that images (and, incidentally, related data such as time series and "signals" from sensors such as microphones or range-finders) are one of the best example types of big data, and one might say that distributed big data analysis of images is inspired by biological systems. After all, many of us perform very sophisticated three-dimensional stereo vision processing as a distributed system every time we drive an automobile.

The good news about including imagery as a big data source is that it's not at all as difficult as it once was. Sophisticated libraries are available to interface with Hadoop and other necessary components, such as graph databases, or a messaging component such as Apache Kafka. Low-level libraries such as OpenCV or BoofCV can provide image processing primitives, if necessary. Writing code is compact and easy. For example, we can write a simple, scrollable image viewer with the following Java class (shown in Listing 1-1).

Listing 1-1. Hello image world: Java code for an image visualizer stub as shown in Figure 1-5

```java
package com.kildane.iabt;
import java.awt.image.RenderedImage;
import java.io.File;
import java.io.IOException;

import javax.media.jai.JAI;
import javax.imageio.ImageIO;
import javax.media.jai.PlanarImage;
import javax.media.jai.widget.ScrollingImagePanel;
import javax.swing.JFrame;

/**
 * Hello IABT world!
 * The worlds most powerful image processing toolkit (for its size)?
 */
public class App
{
    public static void main(String[] args)
    {
        JAI jai = new JAI();
        RenderedImage image = null;
                try {
                        image = ImageIO.read(new File("/Users/kerryk/Documents/SA1_057_62_
                            hr4.png"));
                } catch (IOException e) {
                        e.printStackTrace();
                }
                if (image == null){ System.out.println("Sorry, the image was null"); return; }
                JFrame f = new JFrame("Image Processing Demo for Pro Hadoop Data Analytics");
        ScrollingImagePanel panel = new ScrollingImagePanel(image, 512, 512);
        f.add(panel);
        f.setSize(512, 512);
        f.setVisible(true);
        System.out.println("Hello IABT World, version of JAI is: " + JAI.getBuildVersion());
    }
}
```

Figure 1-5. *Sophisticated third-party libraries make it easy to build image visualization components in just a few lines of code*

A simple image viewer is just the beginning of an image BDA system, however. There is low-level image processing, feature extraction, transformation into an appropriate data representation for analysis, and finally loading out the results to reports, dashboards, or customized result displays.

We will explore the images as big data (IABD) concept more thoroughly in Chapter 14.

1.10.1 Programming Languages Used

First, a word about programming languages. While Hadoop and its ecosystem were originally written in Java, modern Hadoop subsystems have language bindings for almost every conceivable language, including Scala and Python. This makes it very easy to build the kind of polyglot systems necessary to exploit the useful features of a variety of programming languages, all within one application.

1.10.2 Polyglot Components of the Hadoop Ecosystem

In the modern big data analytical arena, one-language systems are few and far between. While many of the older components and libraries we discuss in this book were primarily written in one programming language (for example, Hadoop itself was written in Java while Apache Spark was primarily written in Scala), BDAs as a rule are a composite of different components, sometimes using Java, Scala, Python, and JavaScript within the same application. These multilingual, modular systems are usually known as polyglot systems.

Modern programmers are used to polyglot systems. Some of the need for a multilingual approach is out of necessity: writing a dashboard for the Internet is appropriate for a language such as JavaScript, for example, although one could write a dashboard using Java Swing in stand-alone or even web mode, under duress. It's all a matter of what is most effective and efficient for the application at hand. In this book, we will embrace the polyglot philosophy, essentially using Java for Hadoop-based components, Scala for Spark-based components, Python and scripting as needed, and JavaScript-based toolkits for the front end, dashboards, and miscellaneous graphics and plotting examples.

1.10.3   Hadoop Ecosystem Structure

While the Hadoop Core provides the bedrock that builds the distributed system functionality, the attached libraries and frameworks known as the "Hadoop ecosystem" provide the useful connections to APIs and functionalities which solve application problems and build distributed systems.

We could visualize the Hadoop ecosystem as a kind of "solar system," the individual components of the ecosystem dependent upon the central Hadoop components, with the Hadoop Core at the center "sun" position, as shown in Figure 1-6. Besides providing management and bookkeeping for the Hadoop cluster itself (for example, Zookeeper and Curator), standard components such as Hive and Pig provide data warehousing, and other ancillary libraries such as Mahout provide standard machine learning algorithm support.

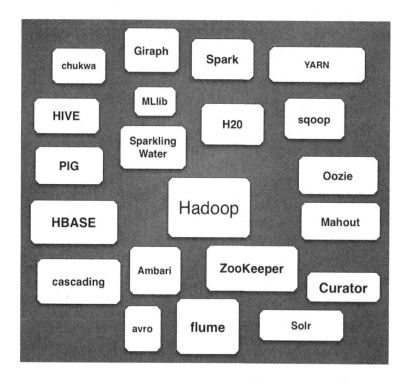

Figure 1-6. *A simplified "solar system" graph of the Hadoop ecosystem*

Apache ZooKeeper (zookeeper.apache.org) is a distributed coordination service for use with a variety of Hadoop- and Spark-based systems. It features a naming service, group membership, locks and carries for distributed synchronization, as well as a highly reliable and centralized registry. ZooKeeper has a hierarchical namespace data model consisting of "znodes." Apache ZooKeeper is open source and is supported by an interesting ancillary component called Apache Curator, a client wrapper for ZooKeeper which is also a rich framework to support ZooKeeper-centric components. We will meet ZooKeeper and Curator again when setting up a configuration to run the Kafka messaging system.

1.11   A Note about "Software Glue" and Frameworks

"Glue" is necessary for any construction project, and software projects are no exception. In fact, some software components, such as the natural language processing (NLP) component Digital Pebble Behemoth (which we will be discussing in detail later) refer to themselves as "glueware." Fortunately, there are also some general purpose integration libraries and packages that are eminently suitable for building BDAs, as shown in Table 1-1.

Table 1-1. *Database types and some examples from industry*

Name	Location	Description
Spring Framework	`http://projects.spring.io/ spring-framework/`	a Java-based framework for application development, has library support for virtually any part of the application development requirements
Apache Tika	`tika.apache.org`	detects and extracts metadata from a wide variety of file types
Apache Camel	`Camel.apache.org`	a "glueware" component which implements enterprise integration patterns (EIPs)
Spring Data	`http://projects.spring.io/ spring-data/`	data access toolkit, tightly coupled to the rest of Spring Framework
Behemoth	`https://github.com/ DigitalPebble/behemoth`	large-scale document analysis "glueware"

To use Apache Camel effectively, it's helpful to know about enterprise integration patterns (EIPs). There are several good books about EIPs and they are especially important for using Apache Camel.[2]

---

[2]The go-to book on Enterprise Integration Patterns (EIPs) is Gregor Hohpe and Bobby Woolf's *Enterprise Integration Patterns: Designing, Building, and Deploying Messaging Solutions*, 2004, Pearson Education Inc. Boston, MA. ISBN 0-321-20068-3.

1.12 Apache Lucene, Solr, and All That: Open Source Search Components

Search components are as important to distributed computing, and especially big data analysis, as the query engine itself. In fact, sometimes a search engine such as Apache Lucene or Apache Solr is a key part of the query engine implementation itself. We can see the interactions between some of these components in Figure 1-7. It turns out that Lucene's Solr components have an ecosystem of their own, albeit not as large in size as the Hadoop ecosystem. Nevertheless, the Lucene ecosystem contains some very relevant software resources for big data analysis. Besides Lucene and Solr, the Lucene ecosystem includes Nutch, an extensible and highly scalable web crawler (nutch.apache.org). The Lily project from NGDATA is a very interesting software framework we can use to leverage HBase, Zookeeper, Solr, and Hadoop seamlessly. Lily clients can use protocols based on Avro to provide connectivity to Lily. Recall that Apache Avro (avro.apache.org) is a data serialization system which provides a compact and fast binary data format with simple integration with dynamic languages.

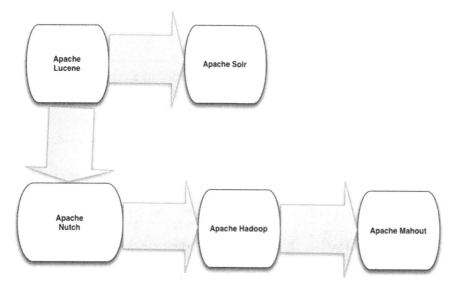

Figure 1-7. *A relationship diagram between Hadoop and other Apache search-related components*

1.13 Architectures for Building Big Data Analytic Systems

Part of the problem when building BDAs is that software development is not really constructing a building. It's just a metaphor, albeit a useful one. When we design a piece of software, we are already using a lot of metaphors and analogies to think about what we're doing. We call it software architecture because it's an analogous process to building a house, and some of the basic principles apply to designing a shopping center as designing a software system.

We want to learn from the history of our technology and not re-invent the wheel or commit the same mistakes as our predecessors. As a result, we have "best practices," software "patterns" and "anti-patterns," methodologies such as Agile or iterative development, and a whole palette of other techniques and strategies. These resources help us achieve quality, reduce cost, and provide effective and manageable solutions for our software requirements.

The "software architecture" metaphor breaks down because of certain realities about software development. If you are building a luxury hotel and you suddenly decide you want to add personal spa rooms or a fireplace to each suite, it's a problem. It's difficult to redesign floor plans, or what brand of carpet to use. There's a heavy penalty for changing your mind. Occasionally we must break out of the building metaphor and take a look at what makes software architecture fundamentally different from its metaphor.

Most of this difference has to do with the dynamic and changeable nature of software itself. Requirements change, data changes, software technologies evolve rapidly. Clients change their minds about what they need and how they need it. Experienced software engineers take this plastic, pliable nature of software for granted, and these realities—the fluid nature of software and of data—impact everything from toolkits to methodologies, particularly the Agile-style methodologies, which assume rapidly changing requirements almost as a matter of course.

These abstract ideas influence our practical software architecture choices. In a nutshell, when designing big data analytical systems, standard architectural principles which have stood the test of time still apply. We can use organizational principles common to any standard Java programming project, for example. We can use enterprise integration patterns (EIPs) to help organize and integrate disparate components throughout our project. And we can continue to use traditional n-tier, client-server, or peer-to-peer principles to organize our systems, if we wish to do so.

As architects, we must also be aware of how distributed systems in general—and Hadoop in particular—change the equation of practical system building. The architect must take into consideration the patterns that apply specifically to Hadoop technologies: for example, mapReduce patterns and anti-patterns. Knowledge is key. So in the next section, we'll tell you what you need to know in order to build effective Hadoop BDAs.

1.14 What You Need to Know

When we wrote this book we had to make some assumptions about you, the reader. We presumed a lot: that you are an experienced programmer and/or architect, that you already know Java, that you know some Hadoop and are acquainted with the Hadoop 2 Core system (including YARN), the Hadoop ecosystem, and that you are used to the basic mechanics of building a Java-style application from scratch. This means that you are familiar with an IDE (such as Eclipse, which we talk about briefly below), that you know about build tools such as Ant and Maven, and that you have a big data analytics problem to solve. We presume you are pretty well-acquainted with the technical issues you want to solve: these include selecting your programming languages, your technology stack, and that you know your data sources, data formats, and data sinks. You may already be familiar with Python and Scala programming languages as well, but we include a quick refresher of these languages—and some thoughts about what they are particularly useful for—in the next chapter. The Hadoop ecosystem has a lot of components and only some of them are relevant to what we'll be discussing, so in Table 1-3 we describe briefly some of the Hadoop ecosystem components we will be using.

It's not just your programming prowess we're making assumptions about. We are also presuming that you are a strategic thinker: that you understand that while software technologies change, evolve, and mutate, sound strategy and methodology (with computer science as well as with any other kind of science) allows you to adapt to new technologies and new problem areas alike. As a consequence of being a strategic thinker, you are interested in data formats.

While data formats are certainly not the most glamorous aspect of big data science, they are one of the most relevant issues to the architect and software engineer, because data sources and their formats dictate, to a certain extent, one very important part of any data pipeline: that initial software component or preprocessor that cleans, verifies, validates, insures security, and ingests data from the data source in anticipation of being processed by the computation engine stage of the pipeline. Hadoop is a critical component of the big data analytics discussed in this book, and to benefit the most from this book, you should have a firm understanding of Hadoop Core and the basic components of the Hadoop ecosystem.

This includes the "classic ecosystem" components such as Hive, Pig, and HBase, as well as glue components such as Apache Camel, Spring Framework, the Spring Data sub-framework, and Apache Kafka messaging system. If you are interested in using relational data sources, a knowledge of JDBC and Spring Framework JDBC as practiced in standard Java programming will be helpful. JDBC has made a comeback in components such as Apache Phoenix (phoenix.apache.org), an interesting combination of relational and Hadoop-based technologies. Phoenix provides low-latency queries over HBase data, using standard SQL syntax in the queries. Phoenix is available as a client-embedded JDBC driver, so an HBase cluster may be accessed with a single line of Java code. Apache Phoenix also provides support for schema definitions, transactions, and metadata.

Table 1-2. *Database types and some examples from industry*

Database Type	Example	Location	Description
Relational	mysql	`mahout.apache.org`	This type of database has been around long enough to acquire sophisticated support frameworks and systems.
Document	Apache Jackrabbit	`jackrabbit.apache.org`	a content repository in Java
Graph	Neo4j	`Neo4j.com`	a multipurpose graph database
File-based	Lucene	`Lucene.apache.org`	statistical, general purpose
Hybrid	Solr+Camel	`Lucene.apache.org/ solr` , `Camel.apache.org`	Lucene, Solr, and glue together as one

■ **Note**   One of the best references for setting up and effectively using Hadoop is the book *Pro Apache Hadoop*, second edition, by Jason Venner and Sameer Wadkhar, available from Apress Publishing.

Some of the toolkits we will discuss are briefly summarized in Table 1-3.

Table 1-3. *A sampling of BDA components in and used with the Hadoop Ecosystem*

Name	Vendor	Location	Description
Mahout	Apache	`mahout.apache.org`	machine learning for Hadoop
MLlib	Apache	`Spark.apache.org/mllib`	machine learning for Apache Spark
R		`https://www.r-project.org`	statistical, general purpose
Weka	University of Waikato, NZ	`http://www.cs.waikato. ac.nz/ml/weka/`	statistical analysis and data mining (Java based)
H2O	H20	`H2o.ai`	JVM-based machine learning
scikit_learn		`scikit-learn.org`	machine learning in Python
Spark	Apache	`spark.apache.org`	open source cluster-computing framework
Kafka	Apache	`kafka.apache.org`	a distributed messaging system

1.15   Data Visualization and Reporting

Data visualization and reporting may be the last step in a data pipeline architecture, but it is certainly as important as the other stages. Data visualization allows the interactive viewing and manipulation of data by the end user of the system. It may be web-based, using RESTful APIs and browsers, mobile devices, or standalone applications designed to run on high-powered graphics displays. Some of the standard libraries for data visualization are shown in Table 1-4.

Table 1-4. *A sampling of front-end components for data visualization*

Name	Location	Description
D3	D3.org	Javascript data visualization
Ggplot2	http://ggplot2.org	data visualization in Python
matplotlib	http://matplotlib.org	Python library for basic plotting
Three.js	http://threejs.org	JavaScript library for three-dimensional graphs and plots
Angular JS	http://angularjs.org	toolkit allowing the creation of modular data visualization components using JavaScript. It's especially interesting because AngularJS integrates well with Spring Framework and other pipeline components.

It's pretty straightforward to create a dashboard or front-end user interface using these libraries or similar ones. Most of the advanced JavaScript libraries contain efficient APIs to connect with databases, RESTful web services, or Java/Scala/Python applications.

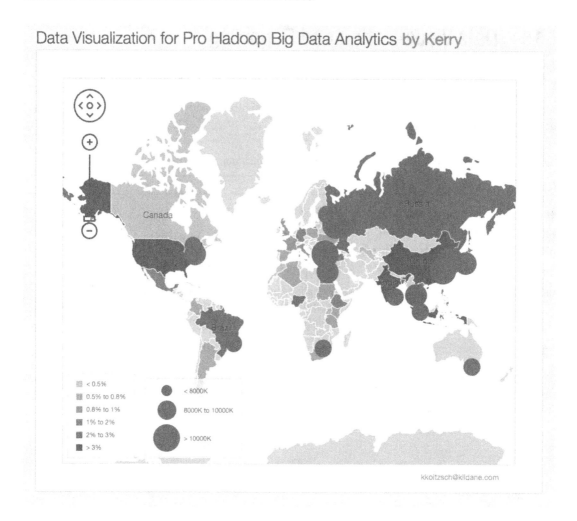

Figure 1-8. *Simple data visualization displayed on a world map, using the DevExpress toolkit*

Big data analysis with Hadoop is something special. For the Hadoop system architect, Hadoop BDA provides and allows the leverage of standard, mainstream architectural patterns, anti-patterns, and strategies. For example, BDAs can be developed using the standard ETL (extract-transform-load) concepts, as well as the architectural principles for developing analytical systems "within the cloud." Standard system modeling techniques still apply, including the "application tier" approach to design.

One example of an application tier design might contain a "service tier" (which provides the "computational engine" or "business logic" of the application) and a data tier (which stores and regulates input and output data, as well as data sources and sinks and an output tier accessed by the system user, which provides content to output devices). This is usually referred to as a "web tier" when content is supplied to a web browser.

ISSUES OF THE PLATFORM

In this book, we express a lot of our examples in a Mac OS X environment. This is by design. The main reason we use the Mac environment is that it seemed the best compromise between a Linux/Unix syntax (which, after all, is where Hadoop lives and breathes) and a development environment on a more modest scale, where a developer could try out some of the ideas shown here without the need for a large Hadoop cluster or even more than a single laptop. This doesn't mean you cannot run Hadoop on a Windows platform in Cygwin or a similar environment if you wish to do so.

Figure 1-9. *A simple data pipeline*

A simple data pipeline is shown in Figure 1-9. In a way, this simple pipeline is the "Hello world" program when thinking about BDAs. It corresponds to the kind of straightforward mainstream ETL (extract-transform-load) process familiar to all data analysts. Successive stages of the pipline transform the previous output contents until the data is emitted to the final data sink or result repository.

1.15.1   Using the Eclipse IDE as a Development Environment

The Eclipse IDE has been around for a long while, and the debate over using Eclipse for modern application development rages on in most development centers that use Java or Scala. There are now many alternatives to Eclipse as an IDE, and you may choose any of these to try out and extend the example systems developed in this book. Or you may even use a regular text editor and run the systems from the command line if you wish, as long as you have the most up-to-date version of Apache Maven around. Appendix A shows you how to set up and run the example systems for a variety of IDEs and platforms, including a modern Eclipse environment. Incidentally, Maven is a very effective tool for organizing the modular Java-based components (as well as components implemented in other languages such as Scala or JavaScript) which make up any BDA, and is integrated directly into the Eclipse IDE. Maven is equally effective on the command line to build, test, and run BDAs.

We have found the Eclipse IDE to be particularly valuable when developing some of the hybrid application examples discussed in this book, but this can be a matter of individual taste. Please feel free to import the examples into your IDE of choice.

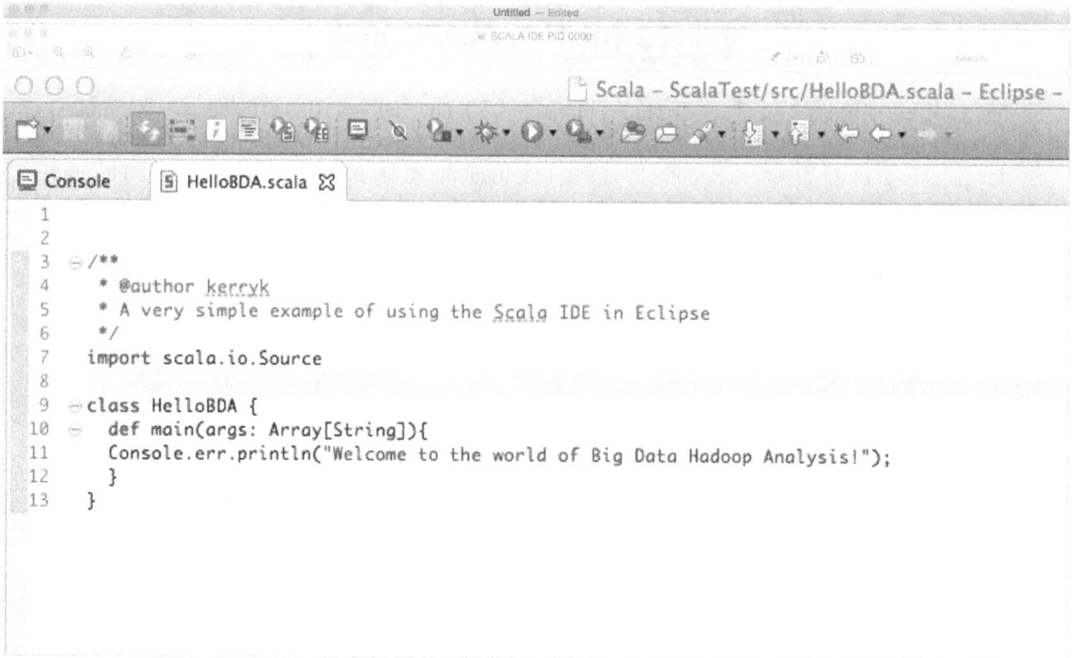

```
Console    HelloBDA.scala ⊠
1
2
3   /**
4    * @author kerryk
5    * A very simple example of using the Scala IDE in Eclipse
6    */
7   import scala.io.Source
8
9   class HelloBDA {
10    def main(args: Array[String]){
11      Console.err.println("Welcome to the world of Big Data Hadoop Analysis!");
12    }
13   }
```

Figure 1-10. *A useful IDE for development : Eclipse IDE with Maven and Scala built in*

DATA SOURCES AND APPLICATION DEVELOPMENT

In mainstream application development—most of the time—we only encounter a few basic types of data sources: relational, various file formats (including raw unstructured text), comma-separated values, or even images (perhaps streamed data or even something more exotic like the export from a graph database such as Neo4j). In the world of big data analysis, signals, images, and non-structured data of many kinds may be used. These may include spatial or GPS information, timestamps from sensors, and a variety of other data types, metadata, and data formats. In this book, particularly in the examples, we will expose you to a wide variety of common as well as exotic data formats, and provide hints on how to do standard ETL operations on the data. When appropriate, we will discuss data validation, compression, and conversion from one data format into another, as needed.

1.15.2   What This Book Is Not

Now that we have given some attention to what this book is about, we must now examine what it is not.

This book is not an introduction to Apache Hadoop, big data analytical components, or Apache Spark. There are many excellent books already in existence which describe the features and mechanics of "vanilla Hadoop" (directly available from hadoop.apache.org) and its ecosystem, as well as the more recent Apache Spark technologies, which are a replacement for the original map-reduce component of Hadoop, and allow for both batch and in-memory processing.

Throughout the book, we will describe useful Hadoop ecosystem components, particularly those which are relevant to the example systems we will be building throughout the rest of this book. These components are building blocks for our BDAs or Big Data Analysis components, so the book will not be discussing the component functionality in depth. In the case of standard Hadoop-compatible components like Apache Lucene, Solr, or Apache Camel or Spring Framework, books and Internet tutorials abound.

We will also not be discussing methodologies (such as iterative or Agile methodologies) in depth, although these are very important aspects of building big data analytical systems. We hope that the systems we are discussing here will be useful to you regardless of what methodology style you choose.

HOW TO BUILD THE BDA EVALUATION SYSTEM

In this section we give a thumbnail sketch of how to build the BDA evaluation system. When completed successfully, this will give you everything you need to evaluate code and examples discussed in the rest of the book. The individual components have complete installation directions at their respective web sites.

1. Set up your basic development environment if you have not already done so. This includes Java 8.0, Maven, and the Eclipse IDE. For the latest installation instructions for Java, visit oracle.com. Don't forget to set the appropriate environment variables accordingly, such as JAVA_HOME. Download and install Maven (maven.apache.org), and set the M2_HOME environment variable. To make sure Maven has been installed correctly, type mvn –version on the command line. Also type 'which mvn' on the command line to insure the Maven executable is where you think it is.

2. Insure that MySQL is installed. Download the appropriate installation package from www.mysql.com/downloads. Use the sample schema and data included with this book to test the functionality. You should be able to run 'mysql' and 'mysqld'.

3. Install the Hadoop Core system. In the examples in this book we use Hadoop version 2.7.1. If you are on the Mac you can use HomeBrew to install Hadoop, or download from the web site and install according to directions. Set the HADOOP_HOME environment variable in your.bash_profile file.

4. Insure that Apache Spark is installed. Experiment with a single-machine cluster by following the instructions at http://spark.apache.org/docs/latest/spark-standalone.html#installing-spark-standalone-to-a-cluster. Spark is a key component for the evaluation system. Make sure the SPARK_HOME environment variable is set in your.bash_profile file.

Figure 1-11. *Successful installation and run of Apache Spark results in a status page at localhost:8080*

To make sure the Spark system is executing correctly, run the program from the SPARK_HOME directory.

```
./bin/run-example SparkPi 10
```

You will see a result similar to the picture in Figure 1-12.

```
spark-1.5.0 — bash — 148×57

Kerrys-MacBook-Pro:spark-1.5.0 kerrys$ ./bin/run-example SparkPi 10
Using Spark's default log4j profile: org/apache/spark/log4j-defaults.properties
15/10/21 15:54:34 INFO SparkContext: Running Spark version 1.5.0
15/10/21 15:54:34 WARN NativeCodeLoader: Unable to load native-hadoop library for your platform... using builtin-java classes where applicable
15/10/21 15:54:34 INFO SecurityManager: Changing view acls to: kerryk
15/10/21 15:54:34 INFO SecurityManager: Changing modify acls to: kerryk
15/10/21 15:54:34 INFO SecurityManager: SecurityManager: authentication disabled; ui acls disabled; users with view permissions: Set(kerryk); users
with modify permissions: Set(kerryk)
15/10/21 15:54:34 INFO Slf4jLogger: Slf4jLogger started
15/10/21 15:54:34 INFO Remoting: Starting remoting
15/10/21 15:54:35 INFO Remoting: Remoting started; listening on addresses :[akka.tcp://sparkDriver@17.115.177.187:60698]
15/10/21 15:54:35 INFO Utils: Successfully started service 'sparkDriver' on port 60698.
15/10/21 15:54:35 INFO SparkEnv: Registering MapOutputTracker
15/10/21 15:54:35 INFO SparkEnv: Registering BlockManagerMaster
15/10/21 15:54:35 INFO DiskBlockManager: Created local directory at /private/var/folders/kf/6fwdssg903x6hq7y0fdgfhxc0000gn/T/blockmgr-39eb4f0c-c76c-
4eb7-a1fc-252f0b83f891
15/10/21 15:54:35 INFO MemoryStore: MemoryStore started with capacity 530.0 MB
15/10/21 15:54:35 INFO HttpFileServer: HTTP File server directory is /private/var/folders/kf/6fwdssg903x6hq7y0fdgfhxc0000gn/T/spark-33f467aa-19ad-42
ec-a669-859e8e6306c7/httpd-dea80b09-eed7-49b2-89bf-80b64a16656f
15/10/21 15:54:35 INFO HttpServer: Starting HTTP Server
15/10/21 15:54:35 INFO Utils: Successfully started service 'HTTP file server' on port 60700.
15/10/21 15:54:35 INFO SparkEnv: Registering OutputCommitCoordinator
15/10/21 15:54:35 INFO Utils: Successfully started service 'SparkUI' on port 4040.
15/10/21 15:54:35 INFO SparkUI: Started SparkUI at http://17.115.177.187:4040
15/10/21 15:54:35 INFO SparkContext: Added JAR file:/Users/kerryk/Downloads/spark-1.5.0/examples/target/scala-2.10/spark-examples-1.5.0-hadoop2.2.0.
jar at http://17.115.177.187:60700/jars/spark-examples-1.5.0-hadoop2.2.0.jar with timestamp 1445468075442
15/10/21 15:54:35 WARN MetricsSystem: Using default name DAGScheduler for source because spark.app.id is not set.
15/10/21 15:54:35 INFO Executor: Starting executor ID driver on host localhost
15/10/21 15:54:35 INFO Utils: Successfully started service 'org.apache.spark.network.netty.NettyBlockTransferService' on port 60701.
15/10/21 15:54:35 INFO NettyBlockTransferService: Server created on 60701
15/10/21 15:54:35 INFO BlockManagerMaster: Trying to register BlockManager
15/10/21 15:54:35 INFO BlockManagerMasterEndpoint: Registering block manager localhost:60701 with 530.0 MB RAM, BlockManagerId(driver, localhost, 60
701)
15/10/21 15:54:35 INFO BlockManagerMaster: Registered BlockManager
15/10/21 15:54:35 INFO SparkContext: Starting job: reduce at SparkPi.scala:36
15/10/21 15:54:35 INFO DAGScheduler: Got job 0 (reduce at SparkPi.scala:36) with 10 output partitions
15/10/21 15:54:35 INFO DAGScheduler: Final stage: ResultStage 0(reduce at SparkPi.scala:36)
15/10/21 15:54:35 INFO DAGScheduler: Parents of final stage: List()
15/10/21 15:54:35 INFO DAGScheduler: Missing parents: List()
15/10/21 15:54:35 INFO DAGScheduler: Submitting ResultStage 0 (MapPartitionsRDD[1] at map at SparkPi.scala:32), which has no missing parents
15/10/21 15:54:36 INFO MemoryStore: ensureFreeSpace(1888) called with curMem=0, maxMem=555755765
15/10/21 15:54:36 INFO MemoryStore: Block broadcast_0 stored as values in memory (estimated size 1888.0 B, free 530.0 MB)
15/10/21 15:54:36 INFO MemoryStore: ensureFreeSpace(1202) called with curMem=1888, maxMem=555755765
15/10/21 15:54:36 INFO MemoryStore: Block broadcast_0_piece0 stored as bytes in memory (estimated size 1202.0 B, free 530.0 MB)
15/10/21 15:54:36 INFO BlockManagerInfo: Added broadcast_0_piece0 in memory on localhost:60701 (size: 1202.0 B, free: 530.0 MB)
15/10/21 15:54:36 INFO SparkContext: Created broadcast 0 from broadcast at DAGScheduler.scala:861
15/10/21 15:54:36 INFO DAGScheduler: Submitting 10 missing tasks from ResultStage 0 (MapPartitionsRDD[1] at map at SparkPi.scala:32)
15/10/21 15:54:36 INFO TaskSchedulerImpl: Adding task set 0.0 with 10 tasks
15/10/21 15:54:36 INFO TaskSetManager: Starting task 0.0 in stage 0.0 (TID 0, localhost, PROCESS_LOCAL, 2164 bytes)
15/10/21 15:54:36 INFO TaskSetManager: Starting task 1.0 in stage 0.0 (TID 1, localhost, PROCESS_LOCAL, 2164 bytes)
15/10/21 15:54:36 INFO TaskSetManager: Starting task 2.0 in stage 0.0 (TID 2, localhost, PROCESS_LOCAL, 2164 bytes)
15/10/21 15:54:36 INFO TaskSetManager: Starting task 3.0 in stage 0.0 (TID 3, localhost, PROCESS_LOCAL, 2164 bytes)
15/10/21 15:54:36 INFO TaskSetManager: Starting task 4.0 in stage 0.0 (TID 4, localhost, PROCESS_LOCAL, 2164 bytes)
15/10/21 15:54:36 INFO TaskSetManager: Starting task 5.0 in stage 0.0 (TID 5, localhost, PROCESS_LOCAL, 2164 bytes)
15/10/21 15:54:36 INFO TaskSetManager: Starting task 6.0 in stage 0.0 (TID 6, localhost, PROCESS_LOCAL, 2164 bytes)
15/10/21 15:54:36 INFO TaskSetManager: Starting task 7.0 in stage 0.0 (TID 7, localhost, PROCESS_LOCAL, 2164 bytes)
15/10/21 15:54:36 INFO Executor: Running task 6.0 in stage 0.0 (TID 6)
```

Figure 1-12. *To test your Spark installation, run the Spark Pi estimator program. A console view of some expected results.*

5. Install Apache Mahout (mahout.apache.org). This is a very useful distributed analytics toolkit. Set appropriate environment variables including MAHOUT_HOME. Run the Mahout test suite to insure it is installed correctly.

6. Install Apache Kafka (kafka.apache.org). This messaging system will figure prominently in our examples. Chapter 3 lists all the steps necessary to set up and thoroughly exercise the Kafka system.

7. Install your favorite NoSQL and graph databases. These might include Cassandra (Cassandra.apache.org), mongoDB (https://www.mongodb.org/downloads#production), etc. If you are interested in the graph analytics part of this book, Neo4j (http://neo4j.com) is a very popular graph database. Our graph analytics examples are all based on Neo4j. In this book, we use Cassandra as our NoSQL database of choice.

8. Install Apache Solr (lucene.apache.org/solr). Download the Solr server zip file, unzip, and follow additional directions from the README file. This configurable Java-based search component can be seamlessly coupled with Hadoop to provide sophisticated, scalable, and customizable search capabilities, leveraging Hadoop and Spark infrastructure.

9. Install the Scala programming languages and Akka. Make sure that you have a support Scala plug-in in your Eclipse IDE. Make sure Scala and the Scala compiler are installed correctly by typing 'scalac –version' and 'which scala' on the command line.

10. Install Python and IPython. On MacOS systems, Python is already available. You may wish to install the Anaconda system, which provides Python, interactive Python, and many useful libraries all as one package.

11. Install H20 (h2o.ai) and Sparkling Water. Once Apache Spark and Akka are installed, we can install H20 and Sparkling Water components.

12. Install appropriate "glue" components. Spring Framework, Spring Data, Apache Camel, and Apache Tika should be installed. There are already appropriate dependencies for these components in the Maven pom.xml shown in Appendix A. You may wish to install some ancillary components such as SpatialHadoop, distributed Weka for Hadoop, and others.

When you have installed all these components, congratulations. You now have a basic software environment in which you can thoroughly investigate big data analytics systems (BDAs). Using this basic system as a starting point, we are ready to explore the individual modules as well as to write some extensions to the basic BDA functionality provided.

1.16 Summary

In this introductory chapter we looked at the changing landscape of big data, methods to ingest, analyze, store, visualize, and understand the ever-increasing ocean of big data in which we find ourselves. We learned that big data sources are varied and numerous, and that these big data sources pose new and challenging questions for the aspiring big data analyst. One of the major challenges facing the big data analyst today is making a selection between all the libraries and toolkits, technology stacks, and methodologies available for big data analytics.

We also took a brief overview of the Hadoop framework, both core components and associated ecosystem components. In spite of this necessarily brief tour of what Hadoop and its ecosystem can do for us as data analysts, we then explored the architectures and strategies that are available to us, with a mind towards designing and implementing effective Hadoop-based analytical systems, or BDAs. These systems will have the scalability and flexibility to solve a wide spectrum of analytical challenges.

The data analyst has a lot of choices when it comes to selecting big data toolkits, and being able to navigate through the bewildering list of features in order to come up with an effective overall technology stack is key to successful development and deployment. We keep it simple (as simple as possible, that is) by focusing on components which integrate relatively seamlessly with the Hadoop Core and its ecosystem.

Throughout this book we will attempt to prove to you that the design and implementation steps outlined above can result in workable data pipeline architectures and systems suitable for a wide range of domains and problem areas. Because of the flexibility of the systems discussed, we will be able to "swap out" modular components as technology changes. We might find that one machine learning or image processing library is more suitable to use, for example, and we might wish to replace the currently existing application library with one of these. Having a modular design in the first place allows us the freedom of swapping out components easily. We'll see this principle in action when we develop the "image as big data" application example in a later chapter.

In the next chapter, we will take a quick overview and refresher of two of the most popular languages for big data analytics—Scala and Python—and explore application examples where these two languages are particularly useful.

■ ■ ■

CHAPTER 2

A Scala and Python Refresher

This chapter contains a quick review of the Scala and Python programming languages used throughout the book. The material discussed here is primarily aimed at Java/C++ programmers who need a quick review of Scala and Python.

■ **Note** A painless way to install Python is to install the Anaconda Python distribution, available at www.continuum.io/downloads. Anaconda provides many additional Python libraries for math and analytics, including support for Hadoop, Weka, R, and many others.

2.1 Motivation: Selecting the Right Language(s) Defines the Application

Selecting the right programming languages for the right tasks defines the application. In many cases, the choices may seem natural: Java for Hadoop-centric components, Scala for Spark-centric ones. Using Java as the main language of a BDA allows access to Spring Framework, Apache Tika, and Apache Camel to provide "glueware" components. However, strategically (and depending upon the nature of your BDA application) you may need to include other languages and other language bindings. This in turn influences the overall technology stack and the nature of the development process itself. For example, a mobile application might require interfacing with low-level code for the mobile device, possibly including the Erlang language, C++ or C, or others.

Another area in which careful programming language choice is key is in the front-end components for displaying and reporting BDA results. Front-end dashboarding and reporting modules may consist only of JavaScript libraries of varying complexity, if they are web-based. Stand-alone scientific applications, however, may be another story. These may use sophisticated visualization libraries in C, C++, Java, or Python.

Careful control, development, and questioning of the technology stack is very important; but in order to select the technology stack components and their language bindings, we must first compare language features.

© Kerry Koitzsch 2017
K. Koitzsch, *Pro Hadoop Data Analytics*, DOI 10.1007/978-1-4842-1910-2_2

2.1.1   Language Features—a Comparison

We are now going to do a quick comparison of the ten most important features Java, Scala, and Python have to offer us, specifically in terms of developing BDA systems. Each feature we discuss is an essential part of modern programming languages, but has particular usefulness when it comes to BDAs. These useful features (the ones we're mostly concerned with) are:

- standard logical, arithmetic, and control structures. Java, Scala, and Python have much in common as far as fundamental language constructs go.

- object orientation. All three of our languages have an object system, and syntax and semantics vary considerably between Java, Scala, and Python.

- database connectivity. Because the whole point of building a BDA is to establish end-to-end data processing pipelines, efficient handling of the data sources—and the exporting to data sinks—is a key consideration of overall design and technology stack choices.

- functional programming support. Functional programming has always been an important part of distributed application development.

- library support, especially machine learning and statistic library support. A host of different libraries exist written in Java, Scala, or Python. Library and framework selection is one of the most challenging problems facing the BDA designer. Modularity and extensibility of the libraries you select, however, is a key requirement to an effective BDA design. Task-specific libraries, like MLlib for machine learning, are particularly useful but create a dependency on Spark and Scala. These dependencies are particularly important to keep in mind.

- dashboard and front-end connectivity. Usually JavaScript toolkits and libraries (such as AngularJS, D3, and others) are sufficient to build sophisticated dashboards and front-end controls, but—as we will see in the rest of the book—there are exceptions to this, particularly in mobile application development.

- "glueware" connectivity and support. This will include both Java-centric connections as well as connectivity to other libraries and frameworks, even those, like Vowpal Wabbit machine learning library, which are written in C++. We can access VW through web services, or even with a Java-native interface (JNI) support library, if we wish.

- read-eval-print loop support. All modern languages have read-eval-print loops (REPLs) except Java, and this is remedied in the Java 9 specification.

- native, multi-core support, and explicit memory management. This varies considerably between our languages, as we will discuss.

- connectivity with Hadoop, Spark, NoSQL databases and their ecosystems. Tools such as PySpark, Spring Data Hadoop, Apache Camel-neo4j, and many others are used to connect the different components you may require in your BDA.

2.2 Review of Scala

This short review of the Scala language consists of five simple code snippets which highlight a variety of language features that we described in our introductory sections. Scala is particularly interesting because of built-in language features such as type inference, closures, currying, and more. Scala also has a sophisticated object system: each value is an object, every operation a method invocation. Scala is also compatible with Java programs. Modern languages always include support for standard data structures, sets, arrays, and vectors. Scala is no exception, and because Scala has a very close affinity to Java, all the data structures familiar to you from Java programming still apply.

■ **Note** In this book we will be discussing Scala version 2.11.7. Type 'scala –version' on the command line to check your installed version of Scala. You may also check your Scala compiler version by typing 'scalac –version' on the command line.

2.2.1 Scala and its Interactive Shell

Let's start with a simple implementation of the quicksort algorithm, and follow that up by testing the routine in the Scala interactive shell. You can see that Listing 2-1 is a simple declarative style Scala program using recursion. If you were to throw the code into your interactive Scala shell, you would see the result shown in Figure y.y. Java programmers can immediately see the similarity between Java and Scala: Scala also uses the JVM and works hand-in-hand with Java. Even the "package" and "import" statements are similar, and the organization of code modules by "packages" in Scala is also similar to that of the Java package system.

Please note that, like Java, Scala provides a convenient object-oriented packaging system. You can also define a Runnable "main" method in a similar way to Java, as shown in Listing 2-1.

Listing 2-1. Simple example of a Scala program which can be tried out in the interactive shell

```scala
/** An example of a quicksort implementation, this one uses a functional style. */
object Sorter {
  def sortRoutine(lst: List[Int]): List[Int] = {
    if (lst.length < 2)
      lst
    else {
      val pivel = lst(lst.length / 2)
      sortRoutine(lst.filter(_ < pivel)) :::
          lst.filter(_ == pivel) :::
          sortRoutine(lst.filter(_ > pivel))
    }
  }

  def main(args: Array[String]) {
    val examplelist = List(11,14,100,1,99,5,7)
    println(examplelist)
    println(sortRoutine(examplelist))
  }
}
```

```
●  ⊛  ⊛                          ▦ bin — java — 97×31
Kerrys-MBP:bin kerryk$ scala
Unable to find a $JAVA_HOME at "/usr", continuing with system-provided Java...
Welcome to Scala version 2.11.7 (Java HotSpot(TM) 64-Bit Server VM, Java 1.8.0_60).
Type in expressions to have them evaluated.
Type :help for more information.

scala> object Sorter {
     |    def sortRoutine(lst: List[Int]): List[Int] = {
     |      if (lst.length < 2)
     |        lst
     |      else {
     |        val pivel = lst(lst.length / 2)
     |        sortRoutine(lst.filter(_ < pivel)) :::
     |           lst.filter(_ == pivel) :::
     |           sortRoutine(lst.filter(_ > pivel))
     |      }
     |    }
     |
     |    def main(args: Array[String]) {
     |      val examplelist = List(11,14,100,1,99,5,7)
     |      println(examplelist)
     |      println(sortRoutine(examplelist))
     |    }
     | }
defined object Sorter

scala> Sorter.main(null)
List(11, 14, 100, 1, 99, 5, 7)
List(1, 5, 7, 11, 14, 99, 100)

scala> ▯
```

Figure 2-1.

Listing 2-2. An example of functional programming in Scala

Functional programming in Scala [includes the results from the Scala REPL as well]

```
scala> def closure1(): Int => Int = {
     | val next = 1
     | def addit(x: Int) = x + next
     | addit
     | }
closure1: ()Int => Int

scala> def closure2() = {
     | val y = 2
     | val f = closure1()
     | println(f(100))
     | }
closure2: ()Unit
```

You can easily use Spark in any of the interactive Scala shells, as shown in Listing 2-3.

Listing 2-3. Simple use of Apache Spark in Scala

NOTE: Please make sure the bdasourcedatafile.dat file is present in your HDFS before running.

```
val bdaTextFile = sc.textFile("hdfs://bdasourcedatafile.dat")
val returnedErrors = bdaTextFile.filter(line => line.contains("ERROR"))
// Count all the errors
returnedErrors.count()
// Count errors mentioning 'Pro Hadoop Analytics'
errors.filter(line => line.contains("Pro Hadoop Analytics")).count()
// Fetch the Pro Hadoop Analytics errors as an array of strings...
returnedErrors.filter(line => line.contains("Pro Hadoop Analytics")).collect()
```

Listing 2-4. Scala example 4: using Apache Kafka to do word counting

KafkaWordCount program in Scala
```
package org.apache.spark.examples.streaming

import java.util.HashMap

import org.apache.kafka.clients.producer.{ProducerConfig, KafkaProducer, ProducerRecord}

import org.apache.spark.streaming._
import org.apache.spark.streaming.kafka._
import org.apache.spark.SparkConf

/**
 * Consumes messages from one or more topics in Kafka and does wordcount.
 * Usage: KafkaWordCount <zkQuorum> <group> <topics> <numThreads>
 *   <zkQuorum> is a list of one or more zookeeper servers that make quorum
 *   <group> is the name of kafka consumer group
 *   <topics> is a list of one or more kafka topics to consume from
 *   <numThreads> is the number of threads the kafka consumer should use
 *
 * Example:
 *    `$ bin/run-example \
 *      org.apache.spark.examples.streaming.KafkaWordCount zoo01,zoo02,zoo03 \
 *      my-consumer-group topic1,topic2 1`
 */
object KafkaWordCount {
  def main(args: Array[String]) {
    if (args.length < 4) {
      System.err.println("Usage: KafkaWordCount <zkQuorum> <group> <topics> <numThreads>")
      System.exit(1)
    }

    StreamingExamples.setStreamingLogLevels()

    val Array(zkQuorum, group, topics, numThreads) = args
    val sparkConf = new SparkConf().setAppName("KafkaWordCount")
    val ssc = new StreamingContext(sparkConf, Seconds(2))
    ssc.checkpoint("checkpoint")
```

```scala
    val topicMap = topics.split(",").map((_, numThreads.toInt)).toMap
    val lines = KafkaUtils.createStream(ssc, zkQuorum, group, topicMap).map(_._2)
    val words = lines.flatMap(_.split(" "))
    val wordCounts = words.map(x => (x, 1L))
      .reduceByKeyAndWindow(_ + _, _ - _, Minutes(10), Seconds(2), 2)
    wordCounts.print()

    ssc.start()
    ssc.awaitTermination()
  }
}

// Produces some random words between 1 and 100.
object KafkaWordCountProducer {

  def main(args: Array[String]) {
    if (args.length < 4) {
      System.err.println("Usage: KafkaWordCountProducer <metadataBrokerList> <topic> " +
        "<messagesPerSec> <wordsPerMessage>")
      System.exit(1)
    }

    val Array(brokers, topic, messagesPerSec, wordsPerMessage) = args

    // Zookeeper connection properties
    val props = new HashMap[String, Object]()
    props.put(ProducerConfig.BOOTSTRAP_SERVERS_CONFIG, brokers)
    props.put(ProducerConfig.VALUE_SERIALIZER_CLASS_CONFIG,
      "org.apache.kafka.common.serialization.StringSerializer")
    props.put(ProducerConfig.KEY_SERIALIZER_CLASS_CONFIG,
      "org.apache.kafka.common.serialization.StringSerializer")

    val producer = new KafkaProducer[String, String](props)

    // Send some messages
    while(true) {
      (1 to messagesPerSec.toInt).foreach { messageNum =>
        val str = (1 to wordsPerMessage.toInt).map(x => scala.util.Random.nextInt(10).toString)
          .mkString(" ")

        val message = new ProducerRecord[String, String](topic, null, str)
        producer.send(message)
      }

      Thread.sleep(1000)
    }
  }
}
```

Lazy evaluation is a "call-by-need" strategy implementable in any of our favorite languages. A simple example of a lazy evaluation exercise is shown in Listing 2-5.

Listing 2-5. Lazy evaluation in Scala

```scala
/* Object-oriented lazy evaluation in Scala */

package probdalazy

object lazyLib {

  /** Delay the evaluation of an expression until it is required. */
  def delay[A](value: => A): Susp[A] = new SuspImpl[A](value)

  /** Get the value of a delayed expression. */
  implicit def force[A](s: Susp[A]): A = s()

  /**
   * Data type of suspended computations. (The name froms from ML.)
   */
  abstract class Susp[+A] extends Function0[A]

  /**
   * Implementation of suspended computations, separated from the
   * abstract class so that the type parameter can be invariant.
   */
  class SuspImpl[A](lazyValue: => A) extends Susp[A] {
    private var maybeValue: Option[A] = None

    override def apply() = maybeValue match {
      case None =>
        val value = lazyValue
        maybeValue = Some(value)
        value
          case Some(value) =>
        value
    }

    override def toString() = maybeValue match {
      case None => "Susp(?)"
      case Some(value) => "Susp(" + value + ")"
    }
  }
}

object lazyEvaluation {
  import lazyLib._

  def main(args: Array[String]) = {
    val s: Susp[Int] = delay { println("evaluating..."); 3 }

    println("s     = " + s)       // show that s is unevaluated
```

```
    println("s()    = " + s())    // evaluate s
    println("s      = " + s)       // show that the value is saved
    println("2 + s = " + (2 + s)) // implicit call to force()

    val sl = delay { Some(3) }
    val sl1: Susp[Some[Int]] = sl
    val sl2: Susp[Option[Int]] = sl1   // the type is covariant

    println("sl2    = " + sl2)
    println("sl2() = " + sl2())
    println("sl2    = " + sl2)
  }
}
```

2.3 Review of Python

In this section we provide a very succinct overview of the Python programming language. Python is a particularly useful resource for building BDAs because of its advanced language features and seamless compatibility with Apache Spark. Like Scala and Java, Python has thorough support for all the usual data structure types you would expect. There are many advantages to using the Python programming language for building at least some of the components in a BDA system. Python has become a mainstream development language in a relatively short amount of time, and part of the reason for this is that it's an easy language to learn. The interactive shell allows for quick experimentation and the ability to try out new ideas in a facile way. Many numerical and scientific libraries exist to support Python, and there are many good books and online tutorials to learn the language and its support libraries.

■ **Note** Throughout the book we will be using Python version 2.7.6 and interactive Python (IPython) version 4.0.0. To check the versions of python you have installed, type `python -version` or `ipython -version` respectively on the command line.

■ **Note**   To run database connectivity examples, please keep in mind we are primarily using the MySQL database from Oracle. This means you must download and install the MySQL connector for Python from the Oracle web site, which is located at `https://dev.mysql.com/downloads/connector/python/2.1.html` The connector is easy to install. On the Mac, simply double-click on the dmg file and follow the directions. You can then test connectivity using an interactive Python shell.

```
                                        ↑ kerryk — Python — 133×52
Kerrys-MacBook-Pro:~ kerryk$ ipython
Python 2.7.6 (default, Sep  9 2014, 15:04:36)
Type "copyright", "credits" or "license" for more information.

IPython 4.0.0 -- An enhanced Interactive Python.
?         -> Introduction and overview of IPython's features.
%quickref -> Quick reference.
help      -> Python's own help system.
object?   -> Details about 'object', use 'object??' for extra details.

In [1]: from sqlite3 import dbapi2 as sqlite

In [2]: db_connection = sqlite.connect('sample.db')

In [3]: dir(db_connection)
Out[3]:
['DataError',
 'DatabaseError',
 'Error',
 'IntegrityError',
 'InterfaceError',
 'InternalError',
 'NotSupportedError',
 'OperationalError',
 'ProgrammingError',
 'Warning',
 '__call__',
 '__class__',
 '__delattr__',
 '__doc__',
 '__enter__',
 '__exit__',
 '__format__',
 '__getattribute__',
 '__hash__',
 '__init__',
 '__new__',
 '__reduce__',
 '__reduce_ex__',
 '__repr__',
 '__setattr__',
 '__sizeof__',
 '__str__',
 '__subclasshook__',
 'close',
 'commit',
 'create_aggregate',
 'create_collation',
 'create_function',
 'cursor',
 'execute',
 'executemany',
```

Figure 2-2. *Simple example of an IPython program, showing database connectivity*

A simple example of database connectivity in Python is shown in Listing 2-6. Readers familiar with Java JDBC constructs will see the similarity. This simple example makes a database connection, then closes it. Between the two statements the programmer can access the designated database, define tables, and perform relational queries.

Listing 2-6. Database connectivity code with Python

Database connectivity example in Python: import, connect, and release (close)

```python
import mysql.connector

cnx = mysql.connector.connect(user='admin', password='',
                              host='127.0.0.1',
                              database='test')
cnx.close()
```

Algorithms of all kinds are easily implemented in Python, and there is a wide range of libraries to assist you. Use of recursion and all the standard programming structures are available. A simple example of a recursive program is shown in Listing 2-7.

Listing 2-7. Recursive Python code that flattens a list

A simple Python code example using recursion

```
def FlattenList(a, result=None):
    result = []
    for x in a:
        if isinstance(x, list):
            FlattenList(x, result)
        else:
            result.append(x)
            return result

        FlattenList([ [0, 1, [2, 3] ], [4, 5], 6])
```

Just as with Java and Scala, it's easy to include support packages with the Python "import" statement. A simple example of this is shown in Listing 2-8.

Planning your import lists explicitly is key to keeping a Python program organized and coherent to the development team and others using the Python code.

Listing 2-8. Python code example using time functions

Python example using time functions

```
import time
size_of_vec = 1000
def pure_python_version():
    t1 = time.time()
    X = range(size_of_vec)
    Y = range(size_of_vec)
    Z = []
    for i in range(len(X)):
        Z.append(X[i] + Y[i])
    return time.time() - t1
def numpy_version():
    t1 = time.time()
    X = np.arange(size_of_vec)
    Y = np.arange(size_of_vec)
    Z = X + Y
    return time.time() - t1
t1 = pure_python_version()
t2 = numpy_version()
print(t1, t2)
print("Pro Data Analytics Numpy in this example, is: " + str(t1/t2) + " faster!")
```

The answer returned in IPython will be similar to:

```
Pro Data Analytics Hadoop Numpy in this example, is:  7.75 faster!
```

The NumPy library provides an extension to the python programming language.

Listing 2-9. Python code example 4: Using the NumPy Library

Python example using the NumPy library
```python
import numpy as np
from timeit import Timer
size_of_vec = 1000
def pure_python_version():
    X = range(size_of_vec)
    Y = range(size_of_vec)
    Z = []
    for i in range(len(X)):
        Z.append(X[i] + Y[i])
def numpy_version():
    X = np.arange(size_of_vec)
    Y = np.arange(size_of_vec)
    Z = X + Y
#timer_obj = Timer("x = x + 1", "x = 0")
timer_obj1 = Timer("pure_python_version()", "from __main__ import pure_python_version")
timer_obj2 = Timer("numpy_version()", "from __main__ import numpy_version")
print(timer_obj1.timeit(10))
print(timer_obj2.timeit(10))
```

Listing 2-10 shows an automatic startup file example.

Listing 2-10. Python code example 5: automatic startup behavior in Python

Python example: using a startup file

```python
import os
filename = os.environ.get('PYTHONSTARTUP')
if filename and os.path.isfile(filename):
    with open(filename) as fobj:
        startup_file = fobj.read()
    exec(startup_file)

import site

site.getusersitepackages()
```

2.4 Troubleshoot, Debug, Profile, and Document

Troubleshooting, regardless of what language you are doing it in, involves identifying and solving immediate and serious problems when running your program. Debugging is also troubleshooting, but implies a less serious difficultly, such as an unexpected error condition, logic error, or other unexpected program result. One example of this distinction is a permissions problem. You can't run your program if you don't have execute permissions on a file. You might need to do a 'chmod' command to fix this.

Additionally, we would suggest that troubleshooting is a mental process. Debugging, on the other hand, can be supported with explicit tools for helping you find bugs, logic errors, unexpected conditions, and the like.

2.4.1   Debugging Resources in Python

In Python, the pdb debugger can be loaded by typing:

```
import pdb
import yourmodule
pdb.run ('yourmodule.test()')
```

or you may use pdb with Python directly by typing:

```
python -m pdb yourprogram.py
```

For profiling Python, Robert Kern's very useful line profiler (https://pypi.python.org/pypi/line_profiler/1.0b3) may be installed by typing the following on the command line:

```
sudo pip install line_profiler
```

Successful installation looks like the picture in Figure 2-3.

```
                                    ⬆ kerryk — bash — 124×18
Kerrys-MacBook-Pro:~ kerryk$ sudo pip install line_profiler
Password:
Collecting line-profiler
/Library/Python/2.7/site-packages/pip-7.1.2-py2.7.egg/pip/_vendor/requests/packages/urllib3/util/ssl_.py:90: InsecurePlatfor
mWarning: A true SSLContext object is not available. This prevents urllib3 from configuring SSL appropriately and may cause
certain SSL connections to fail. For more information, see https://urllib3.readthedocs.org/en/latest/security.html#insecurep
latformwarning.
  InsecurePlatformWarning
  Using cached line_profiler-1.0.tar.gz
Installing collected packages: line-profiler
  Running setup.py install for line-profiler
Successfully installed line-profiler-1.0
/Library/Python/2.7/site-packages/pip-7.1.2-py2.7.egg/pip/_vendor/requests/packages/urllib3/util/ssl_.py:90: InsecurePlatfor
mWarning: A true SSLContext object is not available. This prevents urllib3 from configuring SSL appropriately and may cause
certain SSL connections to fail. For more information, see https://urllib3.readthedocs.org/en/latest/security.html#insecurep
latformwarning.
  InsecurePlatformWarning
Kerrys-MacBook-Pro:~ kerryk$ ⬚
```

Figure 2-3. *Successful installation of the line profiler package*

http://www.huyng.com/posts/python-performance-analysis/ has a very good discussion on profiling Python programs.

Install a memory profiler by typing:

```
sudo pip install -U memory_profiler
```

Why not test your profilers by writing a simple Python program to generate primes, a Fibonacci series, or some other small routine of your choice?

```
                                    ⬆ kerryk — Python — 124×43

IPython 4.0.0 -- An enhanced Interactive Python.
?          -> Introduction and overview of IPython's features.
%quickref -> Quick reference.
help       -> Python's own help system.
object?    -> Details about 'object', use 'object??' for extra details.

In [1]: load_ext memory_profiler

In [2]: load_ext line_profiler

In [3]: from primes import getprimes

In [4]: %mprun -f getprimes getprimes(1000)
/Library/Python/2.7/site-packages/memory_profiler.py:75: UserWarning: psutil module not found. memory_profiler will be slow
  warnings.warn("psutil module not found. memory_profiler will be slow")
Filename: primes.py

Line #    Mem usage    Increment   Line Contents
================================================
     2    20.3 MiB     0.0 MiB     def getprimes(n):
     3    20.3 MiB     0.0 MiB         if n==2:
     4                                     return [2]
     5    20.3 MiB     0.0 MiB         elif n<2:
     6                                     return []
     7    20.3 MiB     0.0 MiB         s=range(3,n+1,2)
     8    20.3 MiB     0.0 MiB         mroot = n ** 0.5
     9    20.3 MiB     0.0 MiB         half=(n+1)/2-1
    10    20.3 MiB     0.0 MiB         i=0
    11    20.3 MiB     0.0 MiB         m=3
    12    20.3 MiB     0.0 MiB         while m <= mroot:
    13    20.3 MiB     0.0 MiB             if s[i]:
    14    20.3 MiB     0.0 MiB                 j=(m*m-3)/2
    15    20.3 MiB     0.0 MiB                 s[j]=0
    16    20.3 MiB     0.0 MiB                 while j<half:
    17    20.3 MiB     0.0 MiB                     s[j]=0
    18    20.3 MiB     0.0 MiB                     j+=m
    19    20.3 MiB     0.0 MiB             i=i+1
    20    20.3 MiB     0.0 MiB             m=2*i+3
    21    20.3 MiB     0.0 MiB         return [2]+[x for x in s if x]
('',)

In [5]: ⬚
```

Figure 2-4. *Profiling Python code using memory and line profilers*

2.4.2   Documentation of Python

When documenting Python code, its very helpful to take a look at the documentation style guide from python.org. This can be found at

    https://docs.python.org/devguide/documenting.html.

2.4.3   Debugging Resources in Scala

In this section we'll discuss resources available to help you debug Scala programs. One of the easiest ways to debug programs is simply to install the Scala plug-in within the Eclipse IDE, create and build Scala projects within Eclipse, and debug and run them there as well. For extensive tutorials on how to do this, please refer to http://scala-ide.org.

2.5   Programming Applications and Example

Building a BDA means building a data pipeline processor. While there are many other ways to conceive and build software systems—including the use of methodologies such as Agile, technological concepts such as object orientation, and enterprise integration patterns (EIPs)—a constant is the pipeline concept.

2.6 Summary

In this chapter, we reviewed the Scala and Python programming languages, and compared them with Java. Hadoop is a Java-centric framework while Apache Spark is written in Scala. Most commonly used BDA components typically have language bindings for Java, Scala, and Python, and we discussed some of these components at a high level.

Each of the languages has particular strengths and we were able to touch on some of the appropriate use cases for Java, Scala, and Python.

We reviewed ways to troubleshoot, debug, profile, and document BDA systems, regardless of what language we're coding the BDAs in, and we discussed a variety of plug-ins available for the Eclipse IDE to work with Python and Scala.

In the next chapter, we will be looking at the necessary ingredients for BDA development: the frameworks and libraries necessary to build BDAs using Hadoop and Spark.

2.7 References

Bowles, Michael. Machine Learning in Python: Essential Techniques for Predictive Analysis. Indianapolis, IN : John Wiley and Sons, Inc., 2015.

Hurwitz, Judith S., Kaufman, Marcia, Bowles, Adrian. Cognitive Computing and Big Data Analytics. Indianapolis, IN: John Wiley and Sons, Inc., 2015.

Odersky, Martin, Spoon, Lex, and Venners, Bill. Programming in Scala, Second Edition. Walnut Creek, CA: Artima Press, 2014.

Younker, Jeff. Foundations of Agile Python Development. New York, NY: Apress/Springer-Verlag New York, 2008.

Ziade, Tarek. Expert Python Programming. Birmingham, UK., PACKT Publishing, 2008.

CHAPTER 3

▦ ▦ ▦

Standard Toolkits for Hadoop and Analytics

In this chapter, we take a look at the necessary ingredients for a BDA system: the standard libraries and toolkits most useful for building BDAs. We describe an example system (which we develop throughout the remainder of the book) using standard toolkits from the Hadoop and Spark ecosystems. We also use other analytical toolkits, such as R and Weka, with mainstream development components such as Ant, Maven, npm, pip, Bower, and other system building tools. "Glueware components" such as Apache Camel, Spring Framework, Spring Data, Apache Kafka, Apache Tika, and others can be used to create a Hadoop-based system appropriate for a variety of applications.

▦ **Note**   A successful installation of Hadoop and its associated components is key to evaluating the examples in this book. Doing the Hadoop installation on the Mac in a relatively painless way is described in `http://amodernstory.com/2014/09/23/installing-hadoop-on-mac-osx-yosemite/` in a post titled "Installing Hadoop on the Mac Part I."

3.1   Libraries, Components, and Toolkits: A Survey

No one chapter could describe all the big data analytics components that are out there to assist you in building BDA systems. We can only suggest the categories of components, talk about some typical examples, and expand on these examples in later chapters.

There are an enormous number of libraries which support BDA system building out there. To get an idea of the spectrum of available techniques, consider the components shown in Figure 3-1. This is not an exclusive list of component types, but when you realize that each component type has a variety of toolkits, libraries, languages, and frameworks to choose from, defining the BDA system technology stack can seem overwhelming at first. To overcome this definition problem, system modularity and flexibility are key.

© Kerry Koitzsch 2017
K. Koitzsch, *Pro Hadoop Data Analytics*, DOI 10.1007/978-1-4842-1910-2_3

Figure 3-1. *A whole spectrum of distributed techniques are available for building BDAs*

One of the easiest ways to build a modular BDA system is to use Apache Maven to manage the dependencies and do most of the simple component management for you. Setting up a simple Maven pom.xml file and creating a simple project in Eclipse IDE is a good way to get the evaluation system going. We can start with a simple Maven pom.xml similar to the one shown in Listing 2-1. Please note the only dependencies shown are for the Hadoop Core and Apache Mahout, the machine learning toolkit for Hadoop we discussed in Chapter 1, which we use frequently in the examples. We will extend the Maven pom file to include all the ancillary toolkits we use later in the book. You can add or subtract components as you wish, simply by removing dependencies from the pom.xml file.

Keep in mind that for every technique shown in the diagram, there are several alternatives. For each choice in the technology stack, there are usually convenient Maven dependencies you can add to your evaluation system to check out the functionality, so it's easy to mix and match components. Including the right "glueware" components can make integration of different libraries less painful.

> ■ **Note** The following important environment variables need to be set to use the book examples effectively:
>
> export BDA_HOME="/Users/kerryk/workspace/bdt"

Listing 3-1. A basic pom.xml file for the evaluation system

```
<project xmlns="http://maven.apache.org/POM/4.0.0" xmlns:xsi="http://www.w3.org/2001/
XMLSchema-instance"
  xsi:schemaLocation="http://maven.apache.org/POM/4.0.0 http://maven.apache.org/
maven-v4_0_0.xsd">
  <modelVersion>4.0.0</modelVersion>
  <groupId>com.kildane</groupId>
  <artifactId>bdt</artifactId>
  <packaging>war</packaging>
  <version>0.0.1-SNAPSHOT</version>
  <name>Big Data Toolkit (BDT) Application</name>
  <url>http://maven.apache.org</url>
  <properties>
  <hadoop.version>0.20.2</hadoop.version>
  </properties>
  <dependencies>
    <dependency>
      <groupId>junit</groupId>
      <artifactId>junit</artifactId>
      <version>3.8.1</version>
      <scope>test</scope>
    </dependency>
    <dependency>
        <groupId>org.apache.hadoop</groupId>
        <artifactId>hadoop-core</artifactId>
        <version>${hadoop.version}</version>
</dependency>
<dependency>
        <groupId>org.apache.mahout</groupId>
        <artifactId>mahout-core</artifactId>
        <version>0.9</version>
</dependency>
  </dependencies>
  <build>
    <finalName>BDT</finalName>
  </build>
</project>
```

The easiest way to build a modular BDA system is to use Apache Maven to manage the dependencies and do most of the simple component management for you. Using a simple pom.xml to get your BDA project started is a good way to experiment with modules, lock in your technology stack, and define system functionality—gradually modifying your dependencies and plug-ins as necessary.

Setting up a simple Maven pom.xml file and creating a simple project in Eclipse IDE is an easy way to get the evaluation system going. We can start with a simple Maven pom.xml similar to the one shown in Listing 3-1. Please note the only dependencies shown are for the Hadoop Core and Apache Mahout, the machine learning toolkit for Hadoop we discussed in Chapter 1, which we use frequently in the examples. We will extend the Maven pom file to include all the ancillary toolkits we use later in the book. You can add or subtract components as you wish, simply by removing dependencies from the pom.xml file.

Let's add a rule system to the evaluation system by way of an example. Simply add the appropriate dependencies for the Drools rule system (Google "drools maven dependencies" for most up to date versions of Drools). The complete pom.xml file (building upon our original) is shown in Listing 3-2. We will be leveraging the functionality of JBoss Drools in a complete analytical engine example in Chapter 8. Please note that we supply dependencies to connect the Drools system with Apache Camel as well as Spring Framework for Drools.

Listing 3-2. Add JBoss Drools dependencies to add rule-based support to your analytical engine. A complete example of a Drools use case is in Chapter 8!

```
<project xmlns="http://maven.apache.org/POM/4.0.0" xmlns:xsi="http://www.w3.org/2001/
XMLSchema-instance"
  xsi:schemaLocation="http://maven.apache.org/POM/4.0.0 http://maven.apache.org/
maven-v4_0_0.xsd">
  <modelVersion>4.0.0</modelVersion>
  <groupId>com.kildane</groupId>
  <artifactId>bdt</artifactId>
  <packaging>war</packaging>
  <version>0.0.1-SNAPSHOT</version>
  <name>Big Data Toolkit (BDT) Application, with JBoss Drools Component</name>
  <url>http://maven.apache.org</url>
  <properties>
  <hadoop.version>0.20.2</hadoop.version>
  </properties>
  <dependencies>
    <dependency>
      <groupId>junit</groupId>
      <artifactId>junit</artifactId>
      <version>3.8.1</version>
      <scope>test</scope>
    </dependency>

<!--  add these five dependencies to your BDA project to achieve rule-based support -->
<dependency>
        <groupId>org.drools</groupId>
        <artifactId>drools-core</artifactId>
        <version>6.3.0.Final</version>
</dependency>
<dependency>
        <groupId>org.drools</groupId>
        <artifactId>drools-persistence-jpa</artifactId>
        <version>6.3.0.Final</version>
</dependency>
```

```
<dependency>
        <groupId>org.drools</groupId>
        <artifactId>drools-spring</artifactId>
        <version>6.0.0.Beta2</version>
</dependency>
<dependency>
        <groupId>org.drools</groupId>
        <artifactId>drools-camel</artifactId>
        <version>6.0.0.Beta2</version>
</dependency>
<dependency>
        <groupId>org.drools</groupId>
        <artifactId>drools-jsr94</artifactId>
        <version>6.3.0.Final</version>
</dependency>
    <dependency>
        <groupId>org.apache.hadoop</groupId>
        <artifactId>hadoop-core</artifactId>
        <version>${hadoop.version}</version>
</dependency>
<dependency>
        <groupId>org.apache.mahout</groupId>
        <artifactId>mahout-core</artifactId>
        <version>0.9</version>
</dependency>
  </dependencies>
  <build>
    <finalName>BDT</finalName>
  </build>
</project>
```

3.2   Using Deep Learning with the Evaluation System

DL4j (http://deeplearning4j.org) is an open source-distributed deep learning library for Java and Scala. It is integrated with Hadoop and Spark.

To install:

```
git clone https://github.com/deeplearning4j/dl4j-0.4-examples.git
```

To build the system:

```
cd $DL4J_HOME directory
```

Then:

```
mvn clean install -DskipTests -Dmaven.javadoc.skip=true
```

To verify the dl4j component is running correctly, type:

```
mvn exec:java -Dexec.mainClass="org.deeplearning4j.examples.tsne.TSNEStandardExample"
-Dexec.cleanupDaemonThreads=false
```

You will see textual output similar to that in Listing y.y. if the component is running successfully.

Listing 3-3. Output from the deep learning 4j test routine

```
[INFO] --- exec-maven-plugin:1.4.0:java (default-cli) @ deeplearning4j-examples ---
o.d.e.t.TSNEStandardExample - Load & Vectorize data....
Nov 01, 2015 1:44:49 PM com.github.fommil.jni.JniLoader liberalLoad
INFO: successfully loaded /var/folders/kf/6fwdssg903x6hq7y0fdgfhxc0000gn/T/jniloader54508704
4337083844netlib-native_system-osx-x86_64.jnilib
o.d.e.t.TSNEStandardExample - Build model....
o.d.e.t.TSNEStandardExample - Store TSNE Coordinates for Plotting....
o.d.plot.Tsne - Calculating probabilities of data similarities..
o.d.plot.Tsne - Mean value of sigma 0.00
o.d.plot.Tsne - Cost at iteration 0 was 98.8718490600586
o.d.plot.Tsne - Cost at iteration 1 was 98.8718490600586
o.d.plot.Tsne - Cost at iteration 2 was 98.8718490600586
o.d.plot.Tsne - Cost at iteration 3 was 98.8718490600586
o.d.plot.Tsne - Cost at iteration 4 was 98.8718490600586
o.d.plot.Tsne - Cost at iteration 5 was 98.8718490600586
o.d.plot.Tsne - Cost at iteration 6 was 98.8718490600586
o.d.plot.Tsne - Cost at iteration 7 was 98.8718490600586
o.d.plot.Tsne - Cost at iteration 8 was 98.87185668945312
o.d.plot.Tsne - Cost at iteration 9 was 98.87185668945312
o.d.plot.Tsne - Cost at iteration 10 was 98.87186431884766
......   ......   ......   .......   .....  ......
o.d.plot.Tsne - Cost at iteration 98 was 98.99024963378906
o.d.plot.Tsne - Cost at iteration 99 was 98.99067687988281
[INFO] ------------------------------------------------------------------------
[INFO] BUILD SUCCESS
[INFO] ------------------------------------------------------------------------
[INFO] Total time: 23.075 s
[INFO] Finished at: 2015-11-01T13:45:06-08:00
[INFO] Final Memory: 21M/721M
[INFO] ------------------------------------------------------------------------
```

To use the deeplearning4j component in our evaluation system, we will now require the most extensive changes to our BDA pom file to date. The complete file is shown in Listing 3-4.

Listing 3-4. Complete listing to include deeplearning 4j components

```
<project xmlns="http://maven.apache.org/POM/4.0.0" xmlns:xsi="http://www.w3.org/2001/
XMLSchema-instance"
        xsi:schemaLocation="http://maven.apache.org/POM/4.0.0 http://maven.apache.org/
maven-v4_0_0.xsd">
        <modelVersion>4.0.0</modelVersion>
        <groupId>com.kildane</groupId>
        <artifactId>bdt</artifactId>
        <packaging>war</packaging>
        <version>0.0.1-SNAPSHOT</version>
        <name>Big Data Toolkit (BDT) Application</name>
        <url>http://maven.apache.org</url>
        <properties>
```

```
<!-- new properties for deep learning (dl4j) components -->
      <nd4j.version>0.4-rc3.5</nd4j.version>
      <dl4j.version> 0.4-rc3.4 </dl4j.version>
      <canova.version>0.0.0.11</canova.version>
      <jackson.version>2.5.1</jackson.version>

      <hadoop.version>0.20.2</hadoop.version>
      <mahout.version>0.9</mahout.version>
</properties>
<!-- distribution management for dl4j  -->
<distributionManagement>
      <snapshotRepository>
            <id>sonatype-nexus-snapshots</id>
            <name>Sonatype Nexus snapshot repository</name>
            <url>https://oss.sonatype.org/content/repositories/snapshots</url>
      </snapshotRepository>
      <repository>
            <id>nexus-releases</id>
            <name>Nexus Release Repository</name>
            <url>http://oss.sonatype.org/service/local/staging/deploy/maven2/</url>
      </repository>
</distributionManagement>
<dependencyManagement>
      <dependencies>
            <dependency>
                  <groupId>org.nd4j</groupId>
                  <artifactId>nd4j-jcublas-7.5</artifactId>
                  <version>${nd4j.version}</version>
            </dependency>
      </dependencies>
</dependencyManagement>
<repositories>
      <repository>
            <id>pentaho-releases</id>
            <url>http://repository.pentaho.org/artifactory/repo/</url>
      </repository>
</repositories>
<dependencies>
      <!-- dependencies for dl4j components -->
      <dependency>
            <groupId>org.deeplearning4j</groupId>
            <artifactId>deeplearning4j-nlp</artifactId>
            <version>${dl4j.version}</version>
      </dependency>
      <dependency>
            <groupId>org.deeplearning4j</groupId>
            <artifactId>deeplearning4j-core</artifactId>
            <version>${dl4j.version}</version>
      </dependency>
```

```
<dependency>
        <groupId>org.nd4j</groupId>
        <artifactId>nd4j-x86</artifactId>
        <version>${nd4j.version}</version>
</dependency>
<dependency>
        <groupId>org.jblas</groupId>
        <artifactId>jblas</artifactId>
        <version>1.2.4</version>
</dependency>
<dependency>
        <artifactId>canova-nd4j-image</artifactId>
        <groupId>org.nd4j</groupId>
        <version>${canova.version}</version>
</dependency>
<dependency>
        <groupId>com.fasterxml.jackson.dataformat</groupId>
        <artifactId>jackson-dataformat-yaml</artifactId>
        <version>${jackson.version}</version>
</dependency>

<dependency>
        <groupId>org.apache.solr</groupId>
        <artifactId>solandra</artifactId>
        <version>UNKNOWN</version>
</dependency>
<dependency>
        <groupId>junit</groupId>
        <artifactId>junit</artifactId>
        <version>3.8.1</version>
        <scope>test</scope>
</dependency>
<dependency>
        <groupId>org.apache.hadoop</groupId>
        <artifactId>hadoop-core</artifactId>
        <version>${hadoop.version}</version>
</dependency>
<dependency>
        <groupId>pentaho</groupId>
        <artifactId>mondrian</artifactId>
        <version>3.6.0</version>
</dependency>
<!-- add these five dependencies to your BDA project to achieve rule-based
        support -->
<dependency>
        <groupId>org.drools</groupId>
        <artifactId>drools-core</artifactId>
        <version>6.3.0.Final</version>
</dependency>
```

```xml
<dependency>
        <groupId>org.drools</groupId>
        <artifactId>drools-persistence-jpa</artifactId>
        <version>6.3.0.Final</version>
</dependency>
<dependency>
        <groupId>org.drools</groupId>
        <artifactId>drools-spring</artifactId>
        <version>6.0.0.Beta2</version>
</dependency>
<dependency>
        <groupId>org.apache.spark</groupId>
        <artifactId>spark-streaming_2.10</artifactId>
        <version>1.5.1</version>
</dependency>
<dependency>
        <groupId>org.drools</groupId>
        <artifactId>drools-camel</artifactId>
        <version>6.0.0.Beta2</version>
</dependency>
<dependency>
        <groupId>org.drools</groupId>
        <artifactId>drools-jsr94</artifactId>
        <version>6.3.0.Final</version>
</dependency>

<dependency>
        <groupId>com.github.johnlangford</groupId>
        <artifactId>vw-jni</artifactId>
        <version>8.0.0</version>
</dependency>
<dependency>
        <groupId>org.apache.mahout</groupId>
        <artifactId>mahout-core</artifactId>
        <version>${mahout.version}</version>
</dependency>
<dependency>
        <groupId>org.apache.mahout</groupId>
        <artifactId>mahout-math</artifactId>
        <version>0.11.0</version>
</dependency>
<dependency>
        <groupId>org.apache.mahout</groupId>
        <artifactId>mahout-hdfs</artifactId>
        <version>0.11.0</version>
</dependency>
</dependencies>
<build>
        <finalName>BDT</finalName>
        <plugins>
                <plugin>
```

```
                        <groupId>org.codehaus.mojo</groupId>
                        <artifactId>exec-maven-plugin</artifactId>
                        <version>1.4.0</version>
                        <executions>
                                <execution>
                                        <goals>
                                                <goal>exec</goal>
                                        </goals>
                                </execution>
                        </executions>
                        <configuration>
                                <executable>java</executable>
                        </configuration>
                </plugin>
                <plugin>
                        <groupId>org.apache.maven.plugins</groupId>
                        <artifactId>maven-shade-plugin</artifactId>
                        <version>1.6</version>
                        <configuration>
                                <createDependencyReducedPom>true</
                                createDependencyReducedPom>
                                <filters>
                                        <filter>
                                                <artifact>*:*</artifact>
                                                <excludes>
                                                        <exclude>org/
                                                        datanucleus/**</exclude>
                                                        <exclude>META-INF/*.SF</
                                                        exclude>
                                                        <exclude>META-INF/*.DSA</
                                                        exclude>
                                                        <exclude>META-INF/*.RSA</
                                                        exclude>
                                                </excludes>
                                        </filter>
                                </filters>
                        </configuration>
                        <executions>
                                <execution>
                                        <phase>package</phase>
                                        <goals>
                                                <goal>shade</goal>
                                        </goals>
                                        <configuration>
                                                <transformers>
                                                        <transformer
                                                                implementation="org.
                apache.maven.plugins.shade.resource.AppendingTransformer">
                                                                <resource>reference.
                                                                conf</resource>
                                                        </transformer>
```

```
                                <transformer
                                        implementation="org.
apache.maven.plugins.shade.resource.ServicesResourceTransformer" />
                                <transformer
                                        implementation="org.
apache.maven.plugins.shade.resource.ManifestResourceTransformer">
                                </transformer>
                            </transformers>
                        </configuration>
                    </execution>
                </executions>
            </plugin>
            <plugin>
                <groupId>org.apache.maven.plugins</groupId>
                <artifactId>maven-compiler-plugin</artifactId>
                <configuration>
                    <source>1.7</source>
                    <target>1.7</target>
                </configuration>
            </plugin>
        </plugins>
    </build>
</project>
```

After augmenting your BDA evaluation project to use this pom.xml, perform the maven clean, install, and package tasks to insure your project compiles correctly.

3.3   Use of Spring Framework and Spring Data

Spring Framework (https://spring.io), and its associated framework Spring Data (projects.spring.io/spring-data), are important glueware components, but the Spring frameworks provide a wide variety of functional resources as well. These include security, ORM connectivity, model-view-controller (MVC)-based application development, and more. Spring Framework uses an aspect-oriented programming approach to address cross-cutting concerns, and fully supports a variety of annotations called "stereotypes" within the Java code, minimizing the need for hand-crafted boilerplate.

We will use Spring Framework throughout this book to leverage the sophisticated functional resources it provides, as well as investigating the Spring Data Hadoop component (projects.spring.io/spring-hadoop/), a seamless integration of Hadoop and Spring. In particular, we will use several Spring Framework components in the complete analytical system we develop in Chapter 9.

3.4   Numerical and Statistical Libraries: R, Weka, and Others

In this section, we will discuss R and Weka statistical libraries. R (r-project.org) is an interpreted high-level language developed specifically for statistical analysis. Weka (http://www.cs.waikato.ac.nz/ml/weka) is a powerful statistics library, providing machine learning algorithms for data mining and other analytical tasks. An interesting new development is the Distributed R and Distributed Weka toolkits. Information about DistributedWekaBase and Distributed Weka, by Mark Hall, may be found at

- http://weka.sourceforge.net/packageMetaData/distributedWekaBase/index.html
- http://weka.sourceforge.net/packageMetaData/distributedWekaHadoop/index.html

3.5 OLAP Techniques in Distributed Systems

OLAP (online analytical processing) is another venerable analytic technique—it's been around since the 1970s—that has had a renaissance in the "big data era." Several powerful libraries and frameworks have been developed to support big data OLAP operations. Two of the most interesting of these are Pentaho's Mondrian (`http://community.pentaho.com/projects/mondrian/`) and a new incubator project at Apache, Apache Kylin (`http://kylin.incubator.apache.org`). Pentaho Mondrian provides an open source analytical engine and its own query language, MDX. To add Pentaho Mondrian to your evaluation system, add this repository, and dependency, to your `pom.xml`:

```
<repository>
    <id>pentaho-releases</id>
    <url>http://repository.pentaho.org/artifactory/repo/</url>
  </repository>

<dependency>
  <groupId>pentaho</groupId>
  <artifactId>mondrian</artifactId>
  <version>3.6.0</version>
</dependency>
```

Apache Kylin provides an ANSI SQL interface and multi-dimensional analysis, leveraging Hadoop functionalities. Business intelligence tools such as Tableau (`get.tableau.com`) are supported by Apache Kylin as well.

We will be developing a complete analytical engine example using Apache Kylin to provide OLAP functionality in Chapter 9.

3.6 Hadoop Toolkits for Analysis: Apache Mahout and Friends

Apache Mahout (`mahout.apache.org`) is a machine learning library specifically designed for use with Apache Hadoop and, with more recent versions of Mahout, Apache Spark as well. Like most modern software frameworks, Mahout is coupled with Samsara, an additional component cooperating with Mahout, to provide an advanced math library support for Mahout functionality. Apache Mahout may also be used with compatible libraries like MLlib. More information about high-level functionality can be found in the numerous tutorials and books on Apache Mahout and other Hadoop-based machine learning packages.

Mahout contains many standard algorithms implemented for distributed processing. Some of these algorithms include classification algorithms such as the random forest classification algorithm, an implementation of the muli-layer perceptron neural net classifier, the naïve Bayes classifier, and many other classifier algorithms. These can be used singly or as stages in a data pipeline, or even in parallel with the right configuration setup.

Vowpal Wabbit (`https://github.com/JohnLangford/vowpal_wabbit`) is an open source project initiated at Yahoo! Inc. and continued by Microsoft Research. Some of VW's features include sparse dimension reduction, fast feature lookups, polynomial learning, and cluster parallel learning, all effective techniques to use in our BDA systems. One of the most interesting extensions of VW is the RESTful web interface, which is available at

For a good discussion of Vowpal-Wabbit, and how to set up and run VW correctly, see `http://zinkov.com/posts/2013-08-13-vowpal-tutorial/`.

To install the VW system, you may need to install the `boost` system first.

On Mac OS, type the following three commands (re-`chmod` your `/usr/local/lib` afterwards if you wish):

```
sudo chmod 777 /usr/local/lib
brew install boost
brew link boost

git clone git://github.com/JohnLangford/vowpal_wabbit.git
cd $VW_HOME
make
make test
```

You may also want to investigate the very interesting web interface to VW, available at `https://github.com/eHarmony/vw-webservice`. To install:

```
git clone https://github.com/eHarmony/vw-webservice.git
cd $VW_WEBSERVICE_HOME
mvn clean install package
```

3.7   Visualization in Apache Mahout

Apache Mahout has built-in visualization capabilities for clustering, based on the `java.awt` graphics package. A simple example of a clustering visualization is shown in Figure 3-2. In the visualization technology chapter, we will discuss extensions and alternatives to this basic system with a mind towards providing more advanced visualization features, extending the visualization controls and displays to include an "image as big data" display as well as some Mahout-centric dashboards.

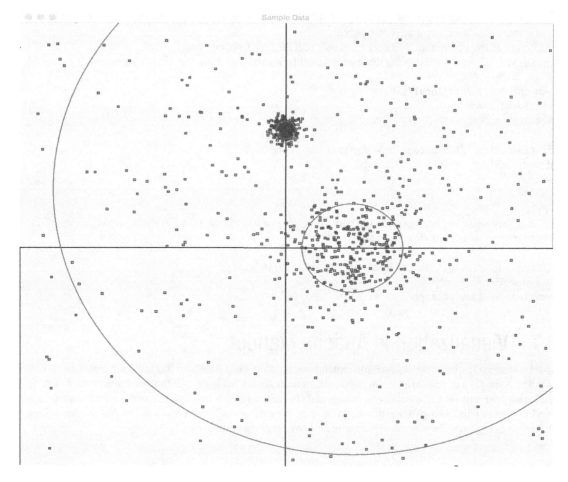

Figure 3-2. *A simple data point visualization using Apache Mahout*

3.8   Apache Spark Libraries and Components

Apache Spark libraries and components are essential to the development of the BDA systems developed in this book. To assist the developer, Spark comes with both Python interactive shell as well as an interactive shell for Scala. As we progress through the book, we will be looking at Apache Spark in detail, as it is one of the most useful alternatives to Hadoop MapReduce technologies. In this section, we will provide a high-level overview of what to expect from the Spark technologies and its ecosystem.

3.8.1   A Variety of Different Shells to Choose From

There are many Python and Scala shells to choose from and in Java 9 we can look forward to a Java-based read-eval-print loop (REPL).

To run the Spark Python shell, type:

```
/bin/pyspark --master spark://server.com:7077 --driver-memory 4g --executor-memory 4g
```

To run the Spark Scala shell, type:

```
./spark-1.2.0/bin/spark-shell --master spark://server.com:7077 --driver-memory 4g
--executor-memory 4g
```

Once you have the `sparkling-water` package installed successfully, you can use the Sparkling shell as shown in Figure 3-4 as your Scala shell. It already has some convenient hooks into Apache Spark for your convenience.

3.8.2   Apache Spark Streaming

Spark Streaming is a fault-tolerant, scalable, and high throughput stream processor.

■ **Note**   Apache Streaming is actively under development. The information about Spark Streaming is constantly subject to change. Refer to `http://spark.apache.org/docs/latest/streaming-programming-guide.html` in order to get the latest information on Apache Streaming. In this book, we primarily refer to the Spark 1.5.1 version.

To add Spark Streaming to your Java project, add this dependency to your `pom.xml` file (get the most recent version parameter to use from the Spark web site):

```xml
<dependency>
    <groupId>org.apache.spark</groupId>
    <artifactId>spark-streaming_2.10</artifactId>
    <version>1.5.1</version>
</dependency>
```

A simplified diagram of the Spark Streaming system is shown in Figure 3-3. Input data streams are processed through the Spark engine and emitted as batches of processed data.

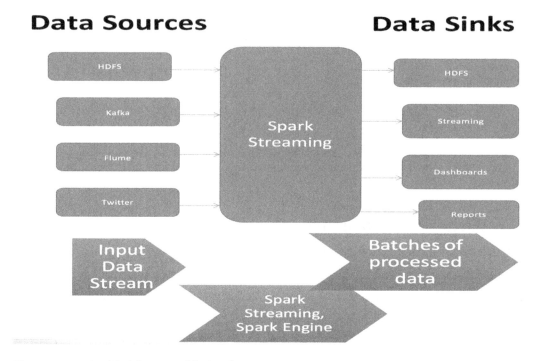

Figure 3-3. *A simplified diagram of the Spark Streaming system*

Spark Streaming is also compatible with Amazon Kinesis (`https://aws.amazon.com/kinesis/`), the AWS data streaming platform.

3.8.3 Sparkling Water and H20 Machine Learning

Sparkling Water (`h20.ai`) is the H20 machine learning toolkit, integrated into Apache Spark. With Sparkling Water, you can use Spark data structures as inputs to H20s algorithms, and there is a Python interface which allows you to use Sparkling Water directly from PyShell.

```
●  ●  ●                        🖿 sparkling-water — bash — 120×38
Kerrys-MacBook-Pro:sparkling-water kerryk$ bin/sparkling-shell

-----
   Spark master (MASTER)       : local-cluster[3,2,2048]
   Spark home    (SPARK_HOME)  : /Users/kerryk/downloads/spark-1.5.0
   H2O build version           : 3.1.0.3118 (master)
   Spark build version         : 1.4.0
----

Unable to find a $JAVA_HOME at "/usr", continuing with system-provided Java...
Unable to find a $JAVA_HOME at "/usr", continuing with system-provided Java...
Java HotSpot(TM) 64-Bit Server VM warning: ignoring option MaxPermSize=384m; support was removed in 8.0
15/10/26 14:52:53 WARN NativeCodeLoader: Unable to load native-hadoop library for your platform... using builtin-java cl
asses where applicable
Welcome to
      ____              __
     / __/__  ___ _____/ /__
    _\ \/ _ \/ _ `/ __/  '_/
   /___/ .__/\_,_/_/ /_/\_\   version 1.5.0
      /_/

Using Scala version 2.10.4 (Java HotSpot(TM) 64-Bit Server VM, Java 1.8.0_60)
Type in expressions to have them evaluated.
Type :help for more information.
15/10/26 14:52:57 WARN MetricsSystem: Using default name DAGScheduler for source because spark.app.id is not set.
Spark context available as sc.
SQL context available as sqlContext.

scala> sc
res0: org.apache.spark.SparkContext = org.apache.spark.SparkContext@4df0d9f8

scala> sqlContext
res1: org.apache.spark.sql.SQLContext = org.apache.spark.sql.SQLContext@13390a96

scala> :quit
Stopping spark context.
15/10/26 14:54:06 WARN QueuedThreadPool: 1 threads could not be stopped
Kerrys-MacBook-Pro:sparkling-water kerryk$ ▯
```

Figure 3-4. *Running the Sparkling Water shell to test your installation*

3.9   Example of Component Use and System Building

In this section we will use the example of the Solandra (Solr + Cassandra) system as a simple example of building a BDA which has all the ingredients necessary to perform big data analytics. In Chapter 1 we had a brief introduction to Solr, the open source, RESTful search engine component which is compatible with both Hadoop and Cassandra NoSQL database. Most of our setup can be done using Maven as shown in Listing 3-4. You'll notice that the pom file listed here is the same as our original project pom file, with dependency additions for Solr, Solandra, and Cassandra components.

1. To download Solandra from the Git source (https://github.com/tjake/Solandra):

    ```
    git clone https://github.com/tjake/Solandra.git
    ```

2. cd to the Solandra directory, and create the JAR file with Maven:

    ```
    cd Solandra
    mvn -DskipTests clean install package
    ```

3. Add the JAR file to your local Maven repository, because there isn't a standard Maven dependency for Solandra yet:

    ```
    mvn install:install-file -Dfile=solandra.jar -DgroupId=solandra
    -DartifactId=solandra -Dpackaging=jar -Dversion=UNKNOWN
    ```

4. Modify your BDA system pom.xml file and add the Solandra dependency:

```
<dependency>
        <groupId>org.apache.solr</groupId>
        <artifactId>solandra</artifactId>
        <version>UNKNOWN</version>
</dependency>
```

5. Test your new BDA pom.xml:

```
cd $BDA_HOME
mvn clean install package
```

BUILDING THE APACHE KAFKA MESSAGING SYSTEM

In this section, we will discuss in detail how to set up and use the Apache Kafka messaging system, an important component of our example BDA framework.

1. Download the Apache Kafka TAR file from `http://kafka.apache.org/downloads.html`

2. Set the KAFKA_HOME environment variable.

3. Unzip file and go to KAFKA_HOME (in this case KAFKA_HOME would be /Users/ kerryk/Downloads/kafka_2.9.1-0.8.2.2).

4. Next, start the ZooKeeper server by typing

    ```
    bin/zookeeper-server-start.sh  config/zookeeper.properties
    ```

5. Once the ZooKeeper service is up and running, type:

    ```
    bin/kafka-server-start.sh config/server.properties
    ```

6. To test topic creation, type:

    ```
    bin/kafka-topics.sh –create –zookeeper localhost:2181 –replication-factor 1 –
    partitions 1 –topic ProHadoopBDAO
    ```

7. To provide a listing of all available topics, type:

    ```
    bin/kafka-topics.sh –list –zookeeper localhost:2181
    ```

    At this stage, the result will be ProHadoopBDAo, the name of the topic you defined in step 5.

8. Send some messages from the console to test the messaging sending functionality. Type:

    ```
    bin/kafka-console-producer.sh –broker-list localhost:9092 –topic ProHadoopBDAO
    ```

    Now type some messages into the console.

9. You can configure a multi broker cluster by modifying the appropriate config files. Check the Apache Kafka documentation for step-by-step processes how to do this.

3.10   Packaging, Testing and Documentation of the Example System

In this section we discuss BDA unit and integration testing. We will discuss Apache Bigtop (bigtop.apache.com) and Apache MRUnit (mrunit.apache.com).

Listing 3-5. Example of Python unit testing from https://docs.python.org/2/library/unittest.html

```python
import unittest

class TestStringMethods(unittest.TestCase):

    def test_upper(self):
        self.assertEqual('foo'.upper(), 'FOO')

    def test_isupper(self):
        self.assertTrue('FOO'.isupper())
        self.assertFalse('Foo'.isupper())

    def test_split(self):
        s = 'hello world'
        self.assertEqual(s.split(), ['hello', 'world'])
        # check that s.split fails when the separator is not a string
        with self.assertRaises(TypeError):
            s.split(2)

if __name__ == '__main__':
    unittest.main()
```

For testing, throughout the book we will use test data sets from http://archive.ics.uci.edu/ml/machine-learning-databases/ as well as the database from Universita de Bologna at http://www.dm.unibo.it/~simoncin/DATA.html. For Python testing, we will be using PyUnit (a Python-based version of the Java unit testing JUnit framework) and pytest (pytest.org), an alternative Python test framework. A simple example of the Python testing component is shown in Listing 3-5.

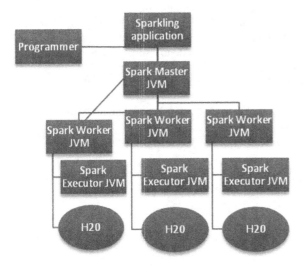

Figure 3-5. *An architecture diagram for the "Sparkling Water" Spark + H20 System*

61

3.11   Summary

In this chapter, we used the first cut of an extensible example system to help motivate our discussion about standard libraries for Hadoop- and Spark-based big data analytics. We also learned that while there are innumerable libraries, frameworks, and toolkits for a wide range of distributed analytic domains, all these components may be tamed by careful use of a good development environment. We chose the Eclipse IDE, Scala and Python plug-in support, and use of the Maven, npm, easy_install, and pip build tools to make our lives easier and to help organize our development process. Using the Maven system alone, we were able to integrate a large number of tools into a simple but powerful image processing module possessing many of the fundamental characteristics of a good BDA data pipelining application.

Throughout this chapter, we have repeatedly returned to our theme of a modular design, showing how a variety of data pipeline systems may be defined and built using the standard ten-step process we discussed in Chapter 1. We also learned about the categories of libraries that are available to help us, including math, statistical, machine learning, image processing, and many others. We discussed in detail how to install and use the Apache Kafka messaging system, an important component we use in our example systems throughout the rest of the book.

There are many language bindings available for these big data Hadoop packages, but we confined our discussion to the Java, Scala, and Python programming languages. You are free to use other language bindings when and if your application demands it.

We did not neglect testing and documentation of our example system. While these components are often seen as "necessary evils," "add-ons," "frills," or "unnecessary," unit and integration testing remain key components of any successful distributed system. We discussed MRUnit and Apache Bigtop as viable testing tools to evaluate BDA systems. Effective testing and documentation lead to effective profiling and optimization, as well as overall system improvement in many other ways.

We not only learned about Hadoop-centric BDA construction using Apache Mahout, but also about using Apache Spark as a fundamental building block, using PySpark, MLlib, H20, and Sparkling Water libraries. Spark technologies for machine learning and BDA construction are now mature and useful ways to leverage powerful machine learning, cognitive computing, and natural language processing libraries to build and extend your own BDA systems.

3.12   References

Giacomelli, Piero. Apache Mahout Cookbook. Birmingham, UK., PACKT Publishing, 2013.

Gupta, Ashish. Learning Apache Mahout Classification. Birmingham, UK., PACKT Publishing, 2015.

Grainger, Trey, and Potter, Timothy. Solr in Action. Shelter Island, NY: Manning Publications, 2014.

Guller, Mohammed. Big Data Analytics with Spark: A Practioner's Guide to Using Spark for Large Scale Data Analysis. Apress/Springer Science+Business Media New York, 2015.

McCandless, Michael, Hatcher, Erik, and Gospodnetic, Otis. Lucene in Action, Second Edition. Shelter Island, NY: Manning Publications, 2010.

Owen, Sean, Anil, Robert, Dunning, Ted, and Friedman, Ellen. Mahout in Action. Shelter Island, NY: Manning Publications, 2011.

Turnbull, Doug, and Berryman, John. Relevant Search: With Applications for Solr and Elasticsearch. Shelter Island, NY: Manning Publications, 2016.

CHAPTER 4

■ ■ ■

Relational, NoSQL, and Graph Databases

In this chapter, we describe the role of databases in distributed big data analysis. Database types include relational databases, document databases, graph databases, and others, which may be used as data sources or sinks in our analytical pipelines. Most of these database types integrate well with Hadoop ecosystem components, as well as with Apache Spark. Connectivity between different kinds of database and Hadoop/ Apache Spark-distributed processing may be provided by "glueware" such as Spring Data or Apache Camel. We describe relational databases, such as MySQL, NoSQL databases such as Cassandra, and graph databases such as Neo4j, and how to integrate them with the Hadoop ecosystem.

There is a spectrum of database types available for you to use, as shown in Figure 4-1. These include flat files (even a CSV file is a kind of database), relational databases such as MySQL and Oracle, key value data stores such as Redis, columnar databases such as HBase (part of the Hadoop ecosystem), as well as more exotic database types such as graph databases (including Neo4J, GraphX, and Giraph)

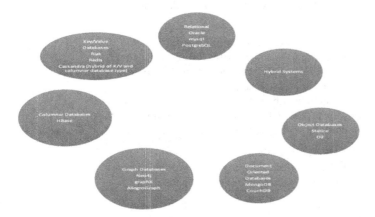

Figure 4-1. *A spectrum of database types*

© Kerry Koitzsch 2017

K. Koitzsch, *Pro Hadoop Data Analytics*, DOI 10.1007/978-1-4842-1910-2_4

We can "abstract out" the concept of different database types as generic data sources, and come up with a common API to connect with, process, and output the content of these data sources. This lets us use different kinds of databases, as needed, in a flexible way. Sometimes it's necessary to adopt a "plug and play" approach for evaluation purposes or to construct proof-of-concept systems. In these instances, it can be convenient to use a NoSQL database such as MongoDB, and compare performance with a Cassandra database or even a graph database component. After evaluation, select the right database for your requirements. Using the appropriate glueware for this purpose, whether it be Apache Camel, Spring Data, or Spring Integration, is key to building a modular system that can be changed rapidly. Much of the glueware code can remain the same, or similar to, the existing code base. Minimum re-work is required if the glueware is selected appropriately.

All database types shown above can be used as distributed system data sources, including relational databases such as MySQL or Oracle. A typical ETL-based processing flow implemented using a relational data source might look like the dataflow shown in Figure 4-2.

1. Cycle Start. The start of the processing cycle is an entry part for the whole system's operation. It's a point of reference for where to start scheduling the processing task, and a place to return to if the system has to undergo a reboot.

2. Reference Data Building. "Reference data" refers to the valid types of data which may be used in individual table fields or the "value" part of key-value pairs.

3. Source Extraction. Retrieve data from the original data sources and do any necessary preprocessing of the data. This might be a preliminary data cleansing or formatting step.

4. Validation Phase. The data is evaluated for consistency.

5. Data Transformation. "Business logic" operations are performed on the data sets to produce an intermediate result.

6. Load into staging tables/data caches or repositories, if used. Staging tables are an intermediate data storage area, which may also be a cache or document database.

7. Report auditing (for business rule compliance, or diagnosis/repair stage). Compute and format report results, export to a displayable format (which may be anything from CSV files to web pages to elaborate interactive dashboard displays). Other forms of report may indicate efficiency of the data process, timings and performance data, system health data, and the like. These ancillary reports support the main reporting task, which is to coherently communicate the results of the data analytics operations on the original data source contents.

8. Publishing to target tables/ repositories. The results so far are exported to the designated output tables or data repositories, which may take a variety of forms including key/value caches, document databases, or even graph databases.

9. Archive back up data. Having a backup strategy is just as important for graph data as traditional data. Replication, validation, and efficient recovery is a must.

10. Log Cycle Status and Errors. We can make use of standard logging constructs, even at the level of Log4j in the Java code, or we may wish to use more sophisticated error logging and reporting if necessary.

Repeat as needed. You can elaborate the individual steps, or specialize to your individual domain problems as required.

4.1   Graph Query Languages : Cypher and Gremlin

Cypher (http://neo4j.com/developer/cypher-query-language/) and Gremlin (http://tinkerpop. incubator.apache.org/gremlin.html) are two of the more well-known graph query languages. Most of the time, graph query languages are designed to be relatively intuitive for programmers with an SQL-style query language background. Graph query languages use nodes, edges, relationships, and patterns to form assertions and queries about data sets modeled as graphs. Refer to Apache TinkerPop's web page (http://tinkerpop.incubator.apache.org) for more information about the Gremlin query language.

To use the new TinkerPop 3 (incubating project at the time this book was written) simply include the following dependency in your pom.xml file:

```
<dependency>
  <groupId>org.apache.tinkerpop</groupId>
  <artifactId>gremlin-core</artifactId>
  <version>3.2.0-incubating</version>
</dependency>
```

Once the dependency is in place in your Java project, you may program to the Java API as shown in Listings 4-1 and 4-2. See the online documentation at: https://neo4j.com/developer/cypher-query-language/ and http://tinkerpop.incubator.apache.org for more information.

4.2   Examples in Cypher

To create a node in Cypher:

```
CREATE (kerry:Person {name:"Kerry"})

RETURN kerry

MATCH (neo:Database {name:"Neo4j"})

MATCH (arubo:Person {name:"Arubo"})

CREATE (anna)-[:FRIEND]->(:Person:Expert {name:"Arubo"})-[:WORKED_WITH]->(neo)
```

To export to a CSV file using cURL:

```
curl -H accept:application/json -H content-type:application/json \
    -d '{"statements":[{"statement":"MATCH (p1:PROFILES)-[:RELATION]-(p2) RETURN ... LIMIT
4"}]}' \
    http://localhost:7474/db/data/transaction/commit \
  | jq -r '(.results[0]) | .columns,.data[].row | @csv'
```

And to time performance, use

```
curl -H accept:application/json -H content-type:application/json \
    -d '{"statements":[{"statement":"MATCH (p1:PROFILES)-[:RELATION]-(p2) RETURN ..."}]}' \
    http://localhost:7474/db/data/transaction/commit \
    | jq -r '(.results[0]) | .columns,.data[].row | @csv' | /dev/null
```

4.3   Examples in Gremlin

The Gremlin graph query language is an alternative to Cypher.

Add a new vertex in the graph

```
g.addVertex([firstName:'Kerry',lastName:'Koitzsch',age:'50']); g.commit();
```

This will require multiple statements. Note how the variables (jdoe and mj) are defined just by assigning them a value from a Gremlin query.

```
jdoe = g.addVertex([firstName:'John',lastName:'Doe',age:'25']);  mj = g.addVertex([firstName
:'Mary',lastName:'Joe',age:'21']); g.addEdge(jdoe,mj,'friend'); g.commit();
```

Add a relation between two existing vertices with id 1 and 2

```
g.addEdge(g.v(1),g.v(2),'coworker'); g.commit();
```

Remove all vertices from the graph:

```
g.V.each{g.removeVertex(it)}
g.commit();
```

Remove all edges from the graph

```
g.E.each{g.removeEdge(it)}
g.commit();
```

Remove all vertices with firstName = 'Kerry'

```
g.V('firstName','Kerry').each{g.removeVertex(it)}
g.commit();
```

Remove a vertex with id 1:

```
g.removeVertex(g.v(1));
g.commit();
```

Remove an edge with id 1

```
g.removeEdge(g.e(1));
g.commit();
```

This is to index the graph with a specific field you may want to search frequently. For example, "myfield"

```
g.createKeyIndex("frequentSearch",Vertex.class);
```

Graphs may also be constructed using the Java API for TinkerPop. In these examples, we will be using the cutting edge version (3-incubating) at the time this book was written.

For a thorough discussion of the TinkerPop system, please see `http://tinkerpop.apache.org`.

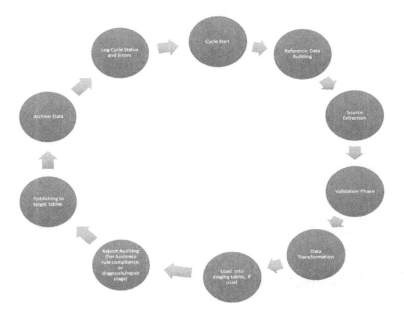

Figure 4-2. *Extract-Transform-Load (ETL) processing lifecycle*

For the purposes of managing data, reference data consists of value sets, or status codes or classification schemas: these are the data objects appropriate for transactions. If we imagine making an ATM withdrawal transaction, for example, we can imagine the associated status codes for such a transaction, such as "Succeeded (S)," "Canceled (CN)," "Funds Not Available (FNA)," "Card Cancelled (CC)," etc.

Reference data is generally uniform, company-wide, and can be either created within a country or by external standardization bodies. Some types of reference data, such as currencies and currency codes, are always standardized. Others, such as the positions of employees within an organization, are less standardized.

Master data and associated transactional data are grouped together as part of transactional records.

Reference data is usually highly standardized, either within the company itself, or by a standardization code supplied by external authorities set up for the purposes of standardization.

Data objects which are relevant to transaction processes are referred to as reference data. These objects may be classification schemas, value sets, or status objects.

Logging cycle status and errors can be as simple as setting the "log levels" in the Java components of the programming and letting the program-based logging do the rest, or the construction of whole systems to do sophisticated logging, monitoring, alerts, and custom reporting. In most cases it is not enough to trust the Java logs alone, of course.

A simple graph database application based on the model-view-controller (MVC) pattern is shown in Figure 4-3. The graph query language can be either Cypher or Gremlin, two graph query languages that we discussed earlier in the chapter.

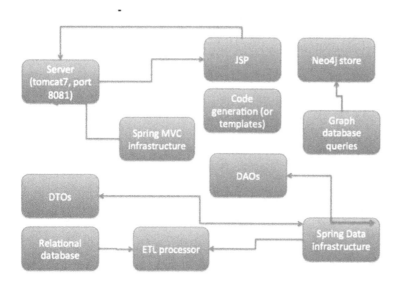

Figure 4-3. *MVC and graph database components*

4.4   Graph Databases: Apache Neo4J

Graph databases are relative newcomers to the NoSQL database arena. One of the most popular and widely used graph databases is the Apache Neo4j package (neo4j.org). Integrating Neo4j graph databases to your distributed analytics application is easy using the Spring Data component for Neo4j (`http://projects.spring.io/spring-data-neo4j/`). Simply make sure the appropriate dependency is present in your pom.xml Maven file:

```
<dependency>
        <groupId>org.springframework.data</groupId>
        <artifactId>spring-data-neo4j</artifactId>
        <version>4.1.1.RELEASE</version>
</dependency>
```

Be sure to remember to supply the correct version number, or make it one of the properties in your pom.xml <properties> tag.

Graph databases can be useful for a number of purposes in a Hadoop-centric system. They can be intermediate result repositories, hold the final results from a computation, or even provide some relatively simple visualization capabilities "out of the box" for dashboarding components, as shown in Figure 4-4.

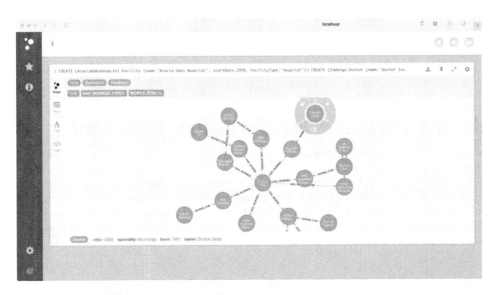

Figure 4-4. *Simple Neo4J data graph visualization*

Let's try a simple load-and-display Neo4j program to get started. The program uses the standard pom.xml included with the "Big Data Analytics Toolkit" software included with this book: This pom.xml includes the necessary dependencies to run our program, which is shown in Listing 4-1.

Listing 4-1. package com.apress.probda.database;

```
import org.neo4j.driver.v1.*;

public class Neo4JExample {

public static void main (String... args){
    // NOTE: on the next line, make sure you have a user defined with the appropriate
    password for your
    // authorization tokens.
    Driver driver = GraphDatabase.driver( "bolt://localhost", AuthTokens.basic( "neo4j",
    "datrosa2016" ) );
    Session session = driver.session();

    session.run( "CREATE (a:Person {name:'Kerry', role:'Programmer'})" );

    StatementResult result = session.run( "MATCH (a:Person) WHERE a.name = 'Kerry' RETURN
    a.name AS name, a.role AS role" );
    while ( result.hasNext() )
    {
        Record record = result.next();
        System.out.println( record.get( "role" ).asString() + " " + record.get("name").
        asString() );
    }
```

```
    System.out.println(".....Simple Neo4J Test is now complete....");
    session.close();
    driver.close();
  }
}
```

4.5   Relational Databases and the Hadoop Ecosystem

Relational databases existed a long time before Hadoop, but they are very compatible with Hadoop, the Hadoop ecosystem, and Apache Spark, too. We can use Spring Data JPA (`http://docs.spring.io/spring-data/jpa/docs/current/reference/html/`) to combine mainstream relational database technology with a distributed environment. The Java Persistence API is a specification (in Java) for managing, accessing, and persisting object-based Java data and a relational database such as MySQL (`dev.mysql.com`). In this section, we will use MySQL as an example of relational database implementation. Many other relational database systems may be used in place of MySQL.

4.6   Hadoop and Unified Analytics (UA) Components

Apache Lens (lens.apache.org) is a new kind of component which provides "unified analytics" (UA) to the Hadoop ecosystem, as shown in Figure 4-5. Unified analytics evolved from the realization that the proliferation of software components, language dialects, and technology stacks made standardization of at least part of the analytics task essential. Unified analytics attempts to standardize data access semantics in the same way that RESTful APIs and semantic web technologies such as RDF (using RDF-REST: `http://liris.cnrs.fr/~pchampin/rdfrest/`) and OWL (`http://owlapi.hets.eu`) provide standardized semantics.

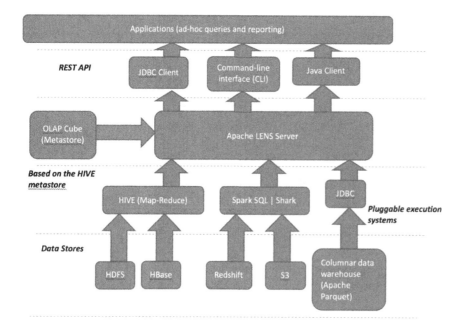

Figure 4-5. *Apache LENS architecture diagram*

As with most of the components we discuss in this book, Apache Lens is easy to install. Download the most recent version for the web site (for our version this was http://www.apache.org/dyn/closer.lua/lens/2.5-beta), expand the zipped TAR file, and run

```
mvn -DskipTests clean package
```

The LENS system, including the Lens UI component, will build, including the Apache Lens UI as shown in Figure 4-6.

Figure 4-6. *Apache LENS installed successfully using Maven on MacOSX*

Log in to Apache Lens by going to the localhost:8784 default Lens web page in any browser. Your login screen will appear as in Figure 4-8.

Run the Lens REPL by typing:

```
./lens-cli.sh
```

You will see a result similar to Figure 4-7. Type 'help' in the interactive shell to see a list of OLAP commands you can try.

Figure 4-7. *Using the Apache Lens REPL*

Figure 4-8. *Apache LENS login page. Use 'admin' for default username and 'admin' for default password.*

Apache Zeppelin (`https://zeppelin.incubator.apache.org`) is a web-based, multipurpose notebook application which enables data ingestion, discovery, and interactive analytics operations. Zeppelin is compatible for use with Scala, SQL, and many other components, languages, and libraries.

```
mvn clean package -Pcassandra-spark-1.5 -Dhadoop.version=2.6.0 -Phadoop-2.6 –DskipTests
```

Figure 4-9. *. Successfully running the Zeppelin browser UI*

Figure 4-10. *Successful Maven build of the Zeppelin notebook*

And then

```
mvn verify
```

Use

```
bin/zeppelin-daemon.sh start
```

to start Zeppelin server, and

```
bin/zeppelin-daemon.sh stop
```

to stop the Zeppelin server. Run the introductory tutorials to test the use of Zeppelin at `https://zeppelin.apache.org/docs/0.6.0/quickstart/tutorial.html`. Zeppelin is particularly useful for interfacing with Apache Spark applications, as well as NoSQL components such as Apache Cassandra.

Data Sources

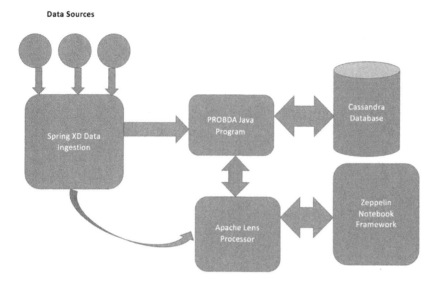

Figure 4-11. *Zeppelin-Lens-Cassandra architecture, with data sources*

OLAP is still alive and well in the Hadoop ecosystem. For example, Apache Kylin (`http://kylin.apache.org`) is an open source OLAP engine for use with Hadoop. Apache Kylin supports distributed analytics, built-in security, and interactive query capabilities, including ANSI SQL support.

Apache Kylin depends on Apache Calcite (`http://incubator.apache.org/projects/calcite.html`) to provide an "SQL core."

To use Apache Calcite, make sure the following dependencies are in your pom.xml file.

```
<dependency>
        <groupId>org.apache.calcite</groupId>
        <artifactId>calcite-core</artifactId>
        <version>1.7.0</version>
</dependency>
```

```
● ● ●                          ⬆ kerryk — bash — 80×24
Last login: Tue May 10 15:06:21 on ttys014
Kerrys-MacBook-Pro:~ kerryk$ curl -L -O http://search.maven.org/remotecontent?fi
lepath=org/hsqldb/sqltool/2.3.2/sqltool-2.3.2.jar
  % Total    % Received % Xferd  Average Speed   Time    Time     Time  Current
                                 Dload  Upload   Total   Spent    Left  Speed
100   161  100   161    0     0    187      0 --:--:-- --:--:-- --:--:--   186
100  143k  100  143k    0     0  54176      0  0:00:02  0:00:02 --:--:-- 89736
Kerrys-MacBook-Pro:~ kerryk$ curl -L -O http://search.maven.org/remotecontent?fi
lepath=org/hsqldb/hsqldb/2.3.2/hsqldb-2.3.2.jar
  % Total    % Received % Xferd  Average Speed   Time    Time     Time  Current
                                 Dload  Upload   Total   Spent    Left  Speed
100   161  100   161    0     0    488      0 --:--:-- --:--:-- --:--:--   487
100 1438k  100 1438k    0     0  71854      0  0:00:20  0:00:20 --:--:-- 309k
Kerrys-MacBook-Pro:~ kerryk$ []
```

Figure 4-12. *HSQLDB installation from the command line*

To install the HSQLDB tools, simply execute

```
curl -L -O http://search.maven.org/remotecontent?filepath=org/hsqldb/sqltool/2.3.2/sqltool-
2.3.2.jar
```

and

```
curl -L -O http://search.maven.org/remotecontent?filepath=org/hsqldb/hsqldb/2.3.2/hsqldb-
2.3.2.jar
```

on the command line. You should see an installation result similar to Figure 4-13. As you can see, Calcite is compatible with many of the databases we have been talking about. Components for use with Cassandra, Spark, and Splunk are available.

```
● ● ●                  apache-calcite-1.7-2.0-src — bash — 88×43
[WARNING]    - com.fasterxml.jackson.core.sym.BytesToNameCanonicalizer
[WARNING]    - com.fasterxml.jackson.core.JsonGenerator$Feature
[WARNING]    - com.fasterxml.jackson.core.io.SegmentedStringWriter
[WARNING]    - com.fasterxml.jackson.core.type.ResolvedType
[WARNING]    - com.fasterxml.jackson.core.TreeNode
[WARNING]    - com.fasterxml.jackson.core.sym.Name
[WARNING]    - com.fasterxml.jackson.core.util.JsonGeneratorDelegate
[WARNING]    - 83 more...
[WARNING] maven-shade-plugin has detected that some .class files
[WARNING] are present in two or more JARs. When this happens, only
[WARNING] one single version of the class is copied in the uberjar.
[WARNING] Usually this is not harmful and you can skeep these
[WARNING] warnings, otherwise try to manually exclude artifacts
[WARNING] based on mvn dependency:tree -Ddetail=true and the above
[WARNING] output
[WARNING] See http://docs.codehaus.org/display/MAVENUSER/Shade+Plugin
[INFO] Replacing /Users/kerryk/Downloads/apache-calcite-1.7-2.0-src/ubenchmark/target/ub
enchmarks.jar with /Users/kerryk/Downloads/apache-calcite-1.7-2.0-src/ubenchmark/target/
calcite-ubenchmark-1.7.0-shaded.jar
[INFO] ------------------------------------------------------------
[INFO] Reactor Summary:
[INFO]
[INFO] Calcite ........................................ SUCCESS [  7.259 s]
[INFO] Calcite Linq4j ................................. SUCCESS [  8.100 s]
[INFO] Calcite Core ................................... SUCCESS [03:27 min]
[INFO] Calcite Cassandra .............................. SUCCESS [  0.859 s]
[INFO] Calcite Examples ............................... SUCCESS [  0.181 s]
[INFO] Calcite Example CSV ............................ SUCCESS [  2.496 s]
[INFO] Calcite Example Function ....................... SUCCESS [  1.770 s]
[INFO] Calcite MongoDB ................................ SUCCESS [  1.361 s]
[INFO] Calcite Piglet ................................. SUCCESS [  2.649 s]
[INFO] Calcite Plus ................................... SUCCESS [  2.940 s]
[INFO] Calcite Spark .................................. SUCCESS [ 14.866 s]
[INFO] Calcite Splunk ................................. SUCCESS [  0.901 s]
[INFO] Calcite Ubenchmark ............................. SUCCESS [  4.121 s]
[INFO] ------------------------------------------------------------
[INFO] BUILD SUCCESS
[INFO] ------------------------------------------------------------
[INFO] Total time: 04:15 min
[INFO] Finished at: 2016-04-22T15:49:14-07:00
[INFO] Final Memory: 142M/1589M
[INFO] ------------------------------------------------------------
Kerrys-MacBook-Pro:apache-calcite-1.7-2.0-src kerryk$ ▯
```

Figure 4-13. *Successful installation of Apache Calcite*

4.7   Summary

In this chapter, we discussed a variety of database types, available software libraries, and how to use the databases in a distributed manner. It should be emphasized that there is a wide spectrum of database technologies and libraries which can be used with Hadoop and Apache Spark. As we discussed, "glueware" such as the Spring Data project, Spring Integration, and Apache Camel, are particularly important when integrating BDA systems with database technologies, as they allow integration of distributed processing technologies with more mainstream database components. The resulting synergy allows the constructed system to leverage relational, NoSQL, and graph technologies to assist with implementation of business logic, data cleansing and validation, reporting, and many other parts of the analytic life cycle.

We talked about two of the most popular graph query languages, Cypher and Gremlin, and looked at some simple examples of these. We took a look at the Gremlin REPL to perform some simple operations there.

When talking about graph databases, we focused on the Neo4j graph database because it is an easy-to-use, full-featured package. Please keep in mind, however, that there are several similar packages which are equally useful, including Apache Giraph (giraph.apache.org),TitanDB (`http://thinkaurelius.github.io/titan/`), OrientDB (`http://orientdb.com/orientdb/`), and Franz's AllegroGraph (`http://franz.com/agraph/allegrograph/`).

In the next chapter, we will discuss distributed data pipelines in more detail—their structure, necessary toolkits, and how to design and implement them.

4.8   References

Hohlpe, Gregor, and Woolf, Bobby. *Enterprise Integration Patterns: Designing, Building, and Deploying Messaging Solutions*. Boston, MA: Addison-Wesley Publishing, 2004.

Ibsen, Claus, and Strachan, James. *Apache Camel in Action*. Shelter Island, NY: Manning Publications, 2010.

Martella, Claudio, Logothetis, Dionysios, Shaposhnik, Roman. *Practical Graph Analytics with Apache Giraph*. New York: Apress Media, 2015.

Pollack, Mark, Gerke, Oliver, Risberg, Thomas, Brisbin, John, and Hunger, Michael. *Spring Data: Modern Data Access for Enterprise Java*. Sebastopol, CA: O'Reilly Media, 2012.

Raj, Sonal. *Neo4J High Performance*. Birmingham, UK: PACKT Publishing, 2015.

Redmond, Eric, and Wilson, Jim R. *Seven Databases in Seven Weeks: A Guide to Modern Databases and the NoSQL Movement*. Raleigh, NC: Pragmatic Programmers, 2012.

Vukotic, Alexa, and Watt, Nicki. *Neo4j in Action*. Shelter Island, NY: Manning Publication, 2015.

■ ■ ■

CHAPTER 5

Data Pipelines and How to Construct Them

In this chapter, we will discuss how to construct basic data pipelines using standard data sources and the Hadoop ecosystem. We provide an end-to-end example of how data sources may be linked and processed using Hadoop and other analytical components, and how this is similar to a standard ETL process. We will develop the ideas presented in this chapter in more detail in Chapter 15.

A NOTE ABOUT THE EXAMPLE SYSTEM STRUCTURE

Since we are going to begin developing the example system in earnest, a note about the package structure of the example system is not out of place here. The basic package structure of the example system developed throughout the book is shown in Figure 5-1, and it's also reproduced in Appendix A. Let's examine what the packages contain and what they do briefly before moving on to data pipeline construction. A brief description of some of the main sub-packages of the Probda system is shown in Figure 5-2.

© Kerry Koitzsch 2017

K. Koitzsch, *Pro Hadoop Data Analytics*, DOI 10.1007/978-1-4842-1910-2_5

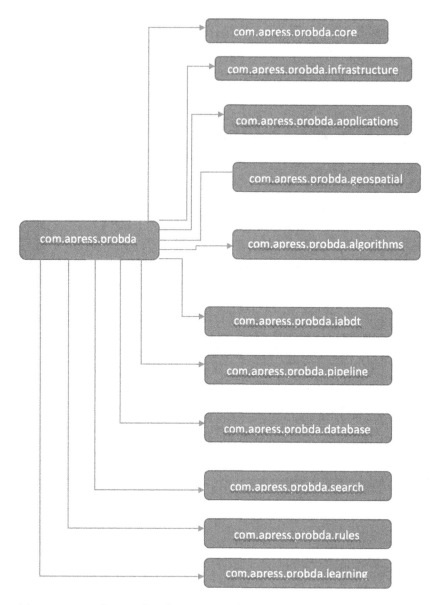

Figure 5-1. *Fundamental package structure for the analytics system*

Package	Description
com.apress.probda	The root package of the Probda system
com.apress.probda.core	Core wrapper classes and fundamental support classes
com.apress.probda.infrastructure	Infrastructure classes and methods
com.apress.probda.applications	The home of the sample applications
com.apress.probda.geospatial	Geospatial support classes
com.apress.probda.algorithms	Algorithm and algorithm support classes
com.apress.probda.iabdt	Image as big data toolkit example application classes
com.apress.probda.database	Support classes for the databases used in this book
com.apress.probda.search	Support classes for the different varieties of search
com.apress.probda.learning	Machine learning and deep learning support and example classes
com.apress.probda.pipeline	Data pipeline classes and support classes

Figure 5-2. *Brief description of the packages in the Probda example system*

In this chapter, we will be concentrating on the classes in the package com.apress.probda.pipeline.

There are five base java classes provided in the code contribution which will enable you to work with reading, transforming, and writing different data sources using a basic data pipelining strategy. See the code contribution notes for more details.

5.1   The Basic Data Pipeline

A basic distributed data pipeline might look like the architecture diagram in Figure 5-3.

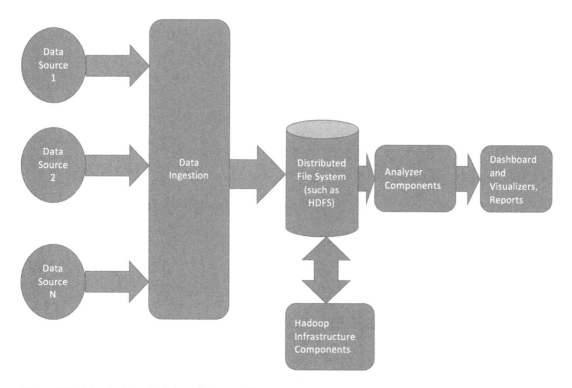

Figure 5-3. *A basic data pipeline architecture diagram*

We can use standard off-the-shelf software components to implement this type of architecture.

We will use Apache Kafka, Beam, Storm, Hadoop, Druid, and Gobblin (formerly Camus) to build our basic pipeline.

5.2   Introduction to Apache Beam

Apache Beam (`http://incubator.apache.org/projects/beam.html`) is a toolkit specifically designed for constructing data pipelines. It has a unified programming model and is designed to be such, since the core of our approach throughout this book is to design and construct distributed data pipelines. Whether using Apache Hadoop, Apache Spark, or Apache Flink as core technologies, Apache Beam fits into the technology stack in a very logical way. At the time this book was written, Apache Beam was an incubating project, so check the web page for its current status.

The key concepts in the Apache Beam programming model are:

- "PCollection": representing a collection of data, which could be bounded or unbounded in size

- "PTransform": representing a computation that transforms input PCollections into output PCollections

- "Pipeline": manages a directed acyclic graph of PTransforms and PCollections that is ready for execution

- "PipelineRunner": specifies where and how the pipeline should execute

These basic elements may be used to construct pipelines with many different topologies, like in the example code in Listing 5-1.

Listing 5-1. Apache Beam test code snippet example

```
static final String[] WORDS_ARRAY = new String[] {
 "probda analytics", "probda", "probda pro analytics",
 "probda one", "three probda", "two probda"};

static final List<String> TEST_WORDS = Arrays.asList(WORDS_ARRAY);

static final String[] WORD_COUNT_ARRAY = new String[] {
 "probda: 6", "one: 1", "pro: 1", "two: 1", "three: 1", "analytics: 2"};

@Test
@Category(RunnableOnService.class)
public void testCountWords() throws Exception {
 Pipeline p = TestPipeline.create();

 PCollection<String> input = p.apply(Create.of(TEST_WORDS).withCoder(StringUtf8Coder.of()));

 PCollection<String> output = input.apply(new CountWords())
 .apply(MapElements.via(new FormatAsTextFn()));

 PAssert.that(output).containsInAnyOrder(WORD_COUNT_ARRAY);
 p.run().waitUntilFinish();
}

cd to contribs/Hadoop and run the Maven file installation
    mvn clean package
```

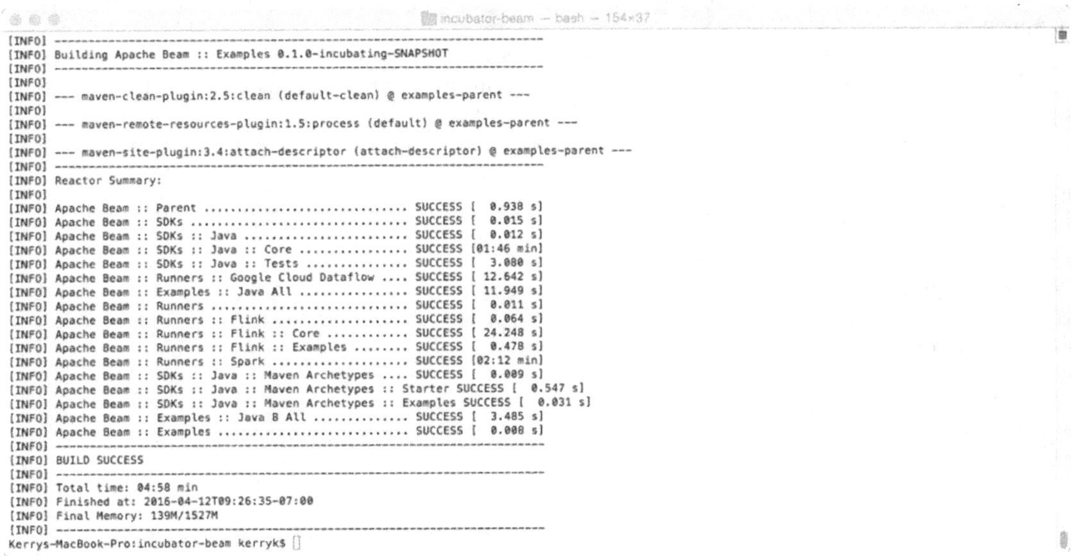

Figure 5-4. *Successful Maven build of Apache Beam, showing the reactor summary*

5.3   Introduction to Apache Falcon

Apache Falcon (`https://falcon.apache.org`) is a feed processing and feed management system aimed at making it easier for end consumers to onboard their feed processing and perform feed management on Hadoop clusters.

Apache Falcon provides the following features:

Apache Falcon (`https://falcon.apache.org`) can be used to process and manage "feeds" on Hadoop clusters, thus providing a system of management which makes it much more straightforward to implement onboarding and establish data flows. It has other useful features, including:

- establishes relationship between various data and processing elements on a Hadoop environment

- feed management services such as feed retention, replications across clusters, archival, etc.

- easy to onboard new workflows/pipelines, with support for late data handling and retry policies

- integration with metastore/catalog such as Hive/HCatalog

- provides notification to end customer based on availability of feed groups (logical group of related feeds, which are likely to be used together)

- enables use cases for local processing in colo and global aggregations

- captures Lineage information for feeds and processes

5.4   Data Sources and Sinks: Using Apache Tika to Construct a Pipeline

Apache Tika (tika.apache.org) is a content analysis toolkit. See the installation instructions for Apache Tika in Appendix A.

Using Apache Tika, almost all mainstream data sources may be used with a distributed data pipeline.

In this example, we will load a special kind of data file, in DBF format, use Apache Tika to process the result, and use a JavaScript visualizer to observe the results of our work.

DBF files are typically used to represent standard database row-oriented data, such as that shown in Listing 5-2.

```
Map: 26 has: 8 entries...
STATION-->Numeric
5203
MAXDAY-->Numeric
20
AV8TOP-->Numeric
9.947581
MONITOR-->Numeric
36203
LAT-->Numeric
```

```
34.107222
LON-->Numeric
-117.273611
X_COORD-->Numeric
474764.37263
Y_COORD-->Numeric
3774078.43207
```

DBF files are typically used to represent standard database row-oriented data, such as that shown in Listing 5-3. A typical method to read DBF files is shown in Listing 5-3.

```
public static List<Map<String, Object>>readDBF(String filename){
        Charset stringCharset = Charset.forName("Cp866");
    List<Map<String,Object>> maps = new ArrayList<Map<String,Object>>();
        try {
        File file = new File(filename);
        DbfReader reader = new DbfReader(file);
        DbfMetadata meta = reader.getMetadata();
        DbfRecord rec = null;
        int i=0;
        while ((rec = reader.read()) != null) {
                rec.setStringCharset(stringCharset);
                Map<String,Object> map = rec.toMap();
                System.out.println("Map: " + i + " has: " + map.size()+ " entries...");

                maps.add(map);
                i++;
        }
        reader.close();
        } catch (IOException e){ e.printStackTrace(); }
        catch (ParseException pe){ pe.printStackTrace(); }
        System.out.println("Read DBF file: " + filename + " , with : " + maps.
                        size()+ " results...");
        return maps
}
```

Gobblin (http://gobblin.readthedocs.io/en/latest/)—formerly known as Camus—is another example of a system based on the "universal analytics paradigm" we talked about earlier.

"Something is missing here: is a universal data ingestion framework for extracting, transforming, and loading large volume of data from a variety of data sources, e.g., databases, rest APIs, FTP/SFTP servers, filers, etc., onto Hadoop. Gobblin handles the common routine tasks required for all data ingestion ETLs, including job/task scheduling, task partitioning, error handling, state management, data quality checking, data publishing, etc. Gobblin ingests data from different data sources in the same execution framework, and manages metadata of different sources all in one place. This, combined with other features such as auto scalability, fault tolerance, data quality assurance, extensibility, and the ability of handling data model evolution, makes Gobblin an easy-to-use, self-serving, and efficient data ingestion framework."

Figure 5-5 shows a successful installation of the Gobblin system.

```
                               ⬚ gobblin -- bash -- 90×55

/gobblin-utility/reports/findbugs/test.xml
:gobblin-utility:test
objc[43508]: Class JavaLaunchHelper is implemented in both /Library/Java/JavaVirtualMachin
es/jdk1.8.0_60.jdk/Contents/Home/bin/java and /Library/Java/JavaVirtualMachines/jdk1.8.0_6
0.jdk/Contents/Home/jre/lib/libinstrument.dylib. One of the two will be used. Which one is
 undefined.
Pass 1: Analyzing classes (39 / 66) - 59% complete                          Pass 1: An
alyzing classes (66 / 66) - 100% complete
javadoc: warning - Error reading file: /Users/kerryk/gobblin/build/gobblin-rest-api/docs/j
avadoc/package-list
Scanning archives (139 / 258)                                         iScanning a
rchives (207 / 258)                              iScanning archives (2
58 / 258)
Pass 2: Analyzing classes (1 / 1) - 100% complete
Done with analysis
FindBugs rule violations were found. See the report at: file:///Users/kerryk/gobblin/build
/gobblin-rest-server/reports/findbugs/test.xml
:gobblin-rest-service:gobblin-rest-server:test
objc[43539]: Class JavaLaunchHelper is implemented in both /Library/Java/JavaVirtualMachin
es/jdk1.8.0_60.jdk/Contents/Home/bin/java and /Library/Java/JavaVirtualMachines/jdk1.8.0_6
0.jdk/Contents/Home/jre/lib/libinstrument.dylib. One of the two will be used. Which one is
 undefined.
1 warning
:gobblin-azkaban:javadocJar
:gobblin-distribution:javadoc UP-TO-DATE
:gobblin-distribution:javadocJar
:gobblin-distribution:assemble
:gobblin-distribution:build
:gobblin-azkaban:assemble
:gobblin-azkaban:build
2 analysis passes to perform
Pass 1: Analyzing classes (39 / 164) - 23% complete                          Pass 1: An
alyzing classes (78 / 164) - 47% complete                       tPass 1: Analyzing cl
asses (156 / 164) - 94% complete                1Pass 1: Analyzing classes (164
 / 164) - 100% complete
Pass 2: Analyzing classes (17 / 17) - 100% complete
Done with analysis
:gobblin-yarn:test
objc[43591]: Class JavaLaunchHelper is implemented in both /Library/Java/JavaVirtualMachin
es/jdk1.8.0_60.jdk/Contents/Home/bin/java and /Library/Java/JavaVirtualMachines/jdk1.8.0_6
0.jdk/Contents/Home/jre/lib/libinstrument.dylib. One of the two will be used. Which one is
 undefined.
:gobblin-rest-service:gobblin-rest-server:check
:gobblin-rest-service:gobblin-rest-server:build
:gobblin-yarn:check
:gobblin-yarn:build
:gobblin-runtime:check
:gobblin-runtime:build
:gobblin-utility:check
:gobblin-utility:build

BUILD SUCCESSFUL

Total time: 8 mins 2.297 secs
Kerrys-MacBook-Pro:gobblin kerryk$ ▯
```

Figure 5-5. *A successful installation of Gobblin*

5.5   Computation and Transformation

Computation and transformation of our data stream can be performed with a small number of simple steps. There are several candidates for this part of the processing pipeline, including Splunk and the commercial software offering Rocana Transform.

We can either use Splunk as a basis for this, or use Rocana Transform. Rocana is a commercial product, so in order to use it you can purchase it or use the free evaluation trial version.

Rocana (https://github.com/scalingdata/rocana-transform-action-plugin) Transform is a configuration-driven transformation library that can be embedded in any JVM-based stream processing or batch processing system such as Spark Streaming, Storm, Flink, or Apache MapReduce.

One of the code contribution examples shows how to build a Rocana transformation engine plug-in, which can perform event data processing within the example system.

In Rocana, a transformation plug-in is made up of two important classes, one based on the Action interface and one based on the ActionBuilder interface, as documented in the code contribution.

5.6 Visualizing and Reporting the Results

Some visualization and reporting can best be done with a notebook-oriented software tool. Most are based on Python—such as Jupyter or Zeppelin. Recall that the Python ecosystem looks something like Figure 5-6. Jupyter and Zeppelin would be under the "Other Packages and Toolboxes" heading, but this does not mean they are not important.

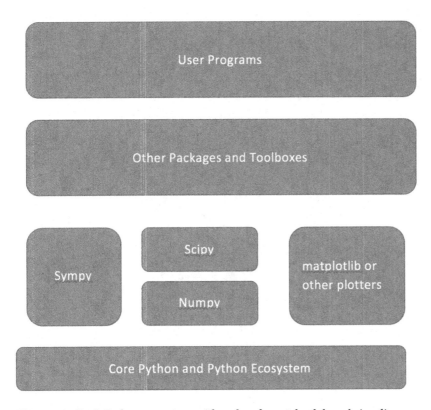

Figure 5-6. *Basic Python ecosystem, with a place for notebook-based visualizers*

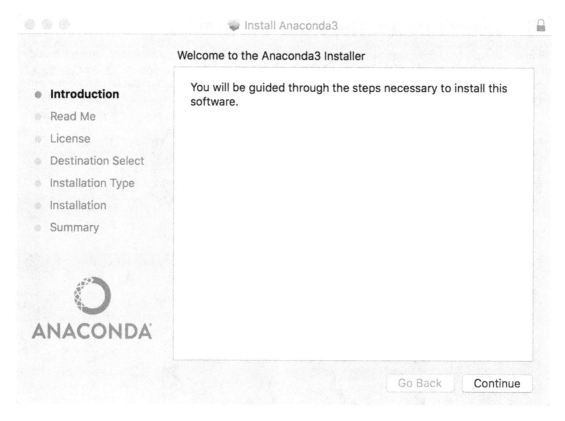

Figure 5-7. *Initial installer diagram for the Anaconda Python system*

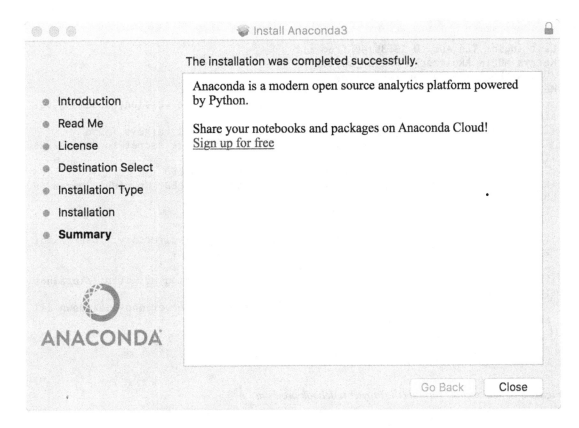

Figure 5-8. *Successful installation of the Anaconda Python system*

We'll be looking at several sophisticated visualization toolkits in chapters to come, but for now let us start out with a quick overview of one of the more popular JavaScript-based toolkits, D3, which can be used to visualize a wide variety of data sources and presentation types. These include geolocations and maps; standard pie, line, and bar charting; tabular reports; and many others (custom presentation types, graph database outputs, and more).

Once Anaconda is working correctly, we can proceed to installing another extremely useful toolkit, TensorFlow. TensorFlow (https://www.tensorflow.org) is a machine learning library which also contains support for a variety of "deep learning" techniques.

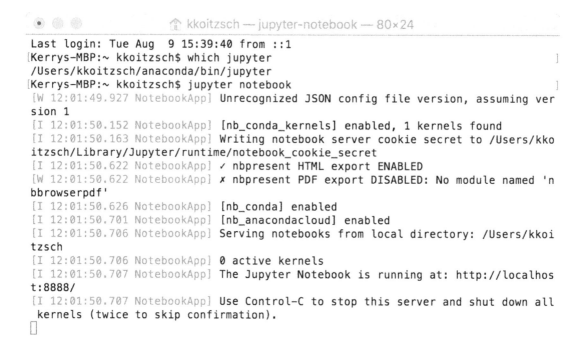

Figure 5-9. *Successfully running the Jupyter notebook program*

```
● ● ●                    ⬆ kkoitzsch — -bash — 80×48
[Kerrys-MBP:~ kkoitzsch$ git clone https://github.com/bokeh/bokeh.git         ]
Cloning into 'bokeh'...
remote: Counting objects: 117730, done.
remote: Compressing objects: 100% (281/281), done.
remote: Total 117730 (delta 156), reused 17 (delta 17), pack-reused 117431
Receiving objects: 100% (117730/117730), 132.66 MiB | 1.63 MiB/s, done.
Resolving deltas: 100% (81821/81821), done.
Checking connectivity... done.
[Kerrys-MBP:~ kkoitzsch$ which conda                                          ]
/Users/kkoitzsch/anaconda/bin/conda
[Kerrys-MBP:~ kkoitzsch$ conda install bokeh                                  ]
Fetching package metadata .......
Solving package specifications: .........

Package plan for installation in environment /Users/kkoitzsch/anaconda:

The following packages will be downloaded:

    package                    |            build
    ---------------------------|-----------------
    anaconda-custom            |           py35_0          3 KB
    conda-env-2.5.2            |           py35_0         27 KB
    conda-4.1.11              |           py35_0        204 KB
    bokeh-0.12.1              |           py35_0        3.3 MB
    ------------------------------------------------------------
                                           Total:        3.5 MB

The following packages will be UPDATED:

    anaconda:  4.1.1-np111py35_0 --> custom-py35_0
    bokeh:     0.12.0-py35_0     --> 0.12.1-py35_0
    conda:     4.1.6-py35_0      --> 4.1.11-py35_0
    conda-env: 2.5.1-py35_0      --> 2.5.2-py35_0

Proceed ([y]/n)? y

Fetching packages ...
anaconda-custo 100% |#############################| Time: 0:00:00    2.19 MB/s
conda-env-2.5. 100% |#############################| Time: 0:00:00  270.90 kB/s
conda-4.1.11-p 100% |#############################| Time: 0:00:00  512.18 kB/s
bokeh-0.12.1-p 100% |#############################| Time: 0:00:02    1.17 MB/s
Extracting packages ...
[      COMPLETE      ]|##############################################| 100%
Unlinking packages ...
[      COMPLETE      ]|##############################################| 100%
Linking packages ...
[      COMPLETE      ]|##############################################| 100%
Kerrys-MBP:~ kkoitzsch$ []
```

Figure 5-10. *Successfully installing Anaconda*

■ **Note**   recall that to build Zeppelin, perform the following steps:

```
mvn clean package -Pcassandra-spark-1.5 -Dhadoop.version=2.6.0 -Phadoop-2.6 -DskipTests
```

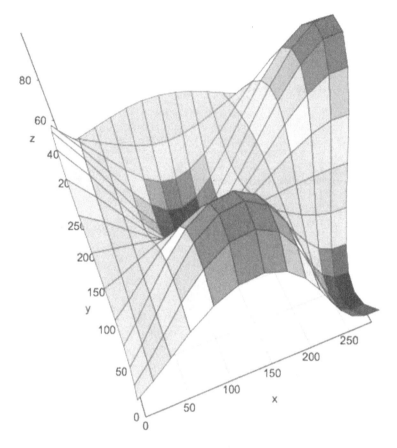

Figure 5-11. *Sophisticated visualizations may be created using the Jupyter visualization feature.*

5.7 Summary

In this chapter, we discussed how to build some basic distributed data pipelines as well as an overview of some of the more effective toolkits, stacks, and strategies to organize and build your data pipeline. Among these were Apache Tika, Gobblin, Spring Integration, and Apache Flink. We also installed Anaconda (which makes the Python development environment much easier to set up and use), as well as an important machine learning library, TensorFlow.

In addition, we took a look at a variety of input and output formats including the ancient but useful DBF format.

In the next chapter, we will discuss advanced search techniques using Lucene and Solr, and introduce some interesting newer extensions of Lucene, such as ElasticSearch.

5.8 References

Lewis, N.D. *Deep Learning Step by Step with Python*. 2016. www.auscov.com

Mattmann, Chris, and Zitting, Jukka. *Tika in Action*. Shelter Island, NY: Manning Publications, 2012.

Zaccone, Giancarlo. *Getting Started with TensorFlow*. Birmingham, UK: PACKT Open Source Publishing, 2016.

CHAPTER 6

■ ■ ■

Advanced Search Techniques with Hadoop, Lucene, and Solr

In this chapter, we describe the structure and use of the Apache Lucene and Solr third-party search engine components, how to use them with Hadoop, and how to develop advanced search capability customized for an analytical application. We will also investigate some newer Lucene-based search frameworks, primarily Elasticsearch, a premier search tool particularly well-suited towards building distributed analytic data pipelines. We will also discuss the extended Lucene/Solr ecosystem and some real-world programming examples of how to use Lucene and Solr in distributed big data analytics applications.

6.1 Introduction to the Lucene/SOLR Ecosystem

As we discussed in the overview of Lucene and Solr in Chapter 1, Apache Lucene (lucene.apache.com) is a key technology to know about when you're building customized search components, and for good reason. It's one of the most venerable Apache components around and has had a long time to mature. In spite of its age, the Lucene/Solr project has been the focus of some interesting new developments in search technology. Lucene and Solr have been merged into one Apache project as of 2010. Some of the main components of the Lucene/Solr ecosystem are shown in Figure 6-1.

K. Koitzsch, *Pro Hadoop Data Analytics*, DOI 10.1007/978-1-4842-1910-2_6

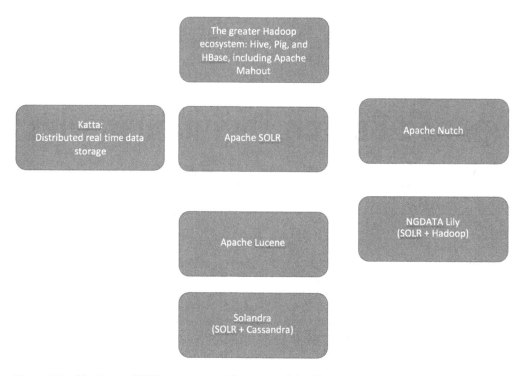

Figure 6-1. *The Lucene/SOLR ecosystem, with some useful additions*

SolrCloud, a new addition to the Lucene/Solr technology stack, allows multicore processing with a RESTful interface. To read more about SolrCloud, visit the information page at `https://cwiki.apache.org/confluence/display/solr/SolrCloud`.

6.2   Lucene Query Syntax

Lucene queries have evolved over the life of the Lucene project to include some sophisticated extensions to the basic query syntax of yesteryear. While Lucene query syntax may change from version to version (and it has evolved considerable since its introduction at Apache in 2001) most of the functionality and search types remain constant, as is shown in Table 6-1.

Table 6-1. *Lucene query types and how to use them*

Type of Search Component	Syntax	Example	Description
free form text	word or "the phrase"	`"to be or not to be"`	either un-quoted words or phrases with double-quotes
keyword search	field name : colon value	`city:Sunnyvale`	field to be searched, a colon, and the string to search for
boosting	term or phrase followed by boost value	`term^3`	Use the caret to provide a new boosting value for a term.
wildcard search	The * symbol can be used for wild carding.	`*kerry`	wild card searches with the '*' or "?" symbol
fuzzy search	Use the tilde to indicate metric distance.	`Hadoop~`	Fuzzy search uses the symbol tilde to indicate closeness using the Levenschein distance metric.
grouping	Normal parentheses provide grouping.	`(java or C)`	Use parentheses to provide sub-queries.
field grouping	Parentheses and colons are used to clarify the query string.	`title:(+gift +"of the magi")`	grouping with field name qualifications, use ordinary parentheses to provide grouping
range search	field name and colon followed by range clause	`startDate:[20020101 TO 20030101] heroes:{Achilles TO Zoroaster}`	Square brackets and the keyword TO allow construction of the range clause, i.e. {Achilles TO Zoroaster}.
proximity Search	term tilde proximity value	`Term~10`	Proximity search uses tilde symbol to indicate "closeness" to the match.

INSTALLING HADOOP, APACHE SOLR, AND NGDATA LILY

In this section we are going to take a brief overview of how to install Hadoop, Lucene/Solr, and NGData's Lily project and suggest some "quick start" techniques to get a Lily installation up and running for development and test purposes.

First, install Hadoop. This is a download, unzip, configure, and run process similar to the many others you have encountered in this book.

When you have successfully installed and configured Hadoop, and and set up the HDFS file system, you should be able to execute some simple Hadoop commands such as

```
hadoop fs -ls /
```

After executing this, you should see a screen similar to the one in Figure 6-2.

```
○ ○ ○                                   bin — bash — 147×39
munishs-macbook-pro:bin kerryk$ which hadoop
/usr/local/bin/hadoop
munishs-macbook-pro:bin kerryk$ hadoop fs -ls /
2016-04-19 15:40:41.341 java[3827:1903] Unable to load realm info from SCDynamicStore
16/04/19 15:40:41 WARN util.NativeCodeLoader: Unable to load native-hadoop library for your platform... using builtin-java classes where applicable
Found 32 items
d--x--x--x   - root wheel         272 2013-05-22 16:23 /.DocumentRevisions-V100
drwxr-xr-x   - root wheel          68 2012-11-16 08:38 /.PKInstallSandboxManager
drwx------   - root wheel         170 2012-11-16 08:44 /.Spotlight-V100
d-wx-wx-wt   - root wheel          68 2013-05-22 16:00 /.Trashes
----------   1 root admin           0 2012-08-22 15:25 /.file
drwx------   - root wheel         204 2016-04-18 13:18 /.fseventsd
-rw-------   1 root wheel     1441792 2016-02-03 07:02 /.hotfiles.btree
drwxr-xr-x   - root wheel          68 2012-06-16 10:41 /.vol
drwxrwxr-x   - root admin        4318 2016-04-14 14:44 /Applications
drwxr-xr-x   - root wheel        2414 2016-01-14 13:16 /Library
drwxr-xr-x   - root wheel          68 2012-08-22 15:25 /Network
drwxr-xr-x   - root wheel         136 2013-01-22 14:16 /System
drwxrwxr-x   - root admin         204 2013-05-22 15:57 /User Information
drwxr-xr-x   - root admin         204 2016-01-14 13:16 /Users
drwxrwxrwt   - root admin         204 2016-04-18 13:19 /Volumes
drwxr-xr-x   - root wheel        1326 2013-06-14 17:13 /bin
drwxrwxr-t   - root admin          68 2012-08-22 15:25 /cores
drwxrwxrwx   - root wheel         238 2014-11-12 14:48 /data
dr-xr-xr-x   - root wheel        4588 2016-04-18 12:55 /dev
drwxr-xr-x   - root wheel        3468 2016-04-18 12:59 /etc
dr-xr-xr-x   - root wheel           1 2016-04-18 12:57 /home
-rw-r--r--   1 root wheel     8194272 2013-05-01 17:57 /mach_kernel
dr-xr-xr-x   - root wheel           1 2016-04-18 12:57 /net
drwxr-xr-x   - root wheel         204 2014-08-20 10:02 /opt
drwxr-xr-x   - root wheel         204 2012-11-16 08:30 /private
drwxr-xr-x   - root wheel        2108 2013-06-14 17:13 /sbin
drwxrwxrwx   - root wheel         170 2014-03-21 15:29 /temp
drwxrwxrwt   - root wheel         442 2016-04-19 15:31 /tmp
drwxr-xr-x   - root wheel         102 2013-10-29 13:57 /user
drwxr-xr-x   - root wheel         476 2013-12-20 11:14 /usr
drwxr-xr-x   - root wheel        1190 2016-04-19 14:44 /var
-rw-r--r--   1 root wheel           5 2016-02-10 15:40 /zookeeper_server.pid
munishs-macbook-pro:bin kerryk$ []
```

Figure 6-2. *Successful test of installation of Hadoop and the Hadoop Distributed File System (HDFS)*

Second, install Solr. This is simply a matter of downloading the zip file at, uncompressing, and cd'ing to the binary file, where you may then start the server immediately, using the command.

A successful installation of Solr can be tested as in Figure 6-3.

```
⚙ ⚙ ⚙                            bin — bash — 80×24
Last login: Tue Apr 19 08:10:50 on ttys009
Kerrys-MBP:~ kerryk$ cd Downloads
Kerrys-MBP:Downloads kerryk$ cd solr-5*
Kerrys-MBP:solr-5.3.0 kerryk$ cd bin
Kerrys-MBP:bin kerryk$ ./solr start
Waiting up to 30 seconds to see Solr running on port 8983 [|]
Started Solr server on port 8983 (pid=31671). Happy searching!

Kerrys-MBP:bin kerryk$ []
```

Figure 6-3. *A successful installation and start of the Solr server*

Third, download NGDATA's Lily project from the github project at `https://github.com/NGDATA/lilyproject`.

Getting Hadoop, Lucene, Solr, and Lily to cooperate in the same software environment can be tricky, so we include some tips on setting up the environment that you may have forgotten.

TIPS ON USING HADOOP WITH SOLR AND LUCENE

1. Make sure you can log in with 'ssh' without password. This is essential for Hadoop to work correctly. It doesn't hurt to exercise your Hadoop installation from time to time, to insure all the moving parts are working correctly. A quick test of Hadoop functionality can be accomplished on the command line with just a few commands. For example:

2. Make sure your environment variables are set correctly, and configure your init files appropriately. This includes such things as your .bash_profile file, if you are on MacOS, for example.

3. Test component interaction frequently. There are a lot of moving parts in distributed systems. Perform individual tests to insure each part is working smoothly.

4. Test interaction in standalone, pseudo-distributed, and full-distributed modes when appropriate. This includes investigating suspicious performance problems, hang-ups, unexpected stalls and errors, and version incompatibilities.

5. Watch out for version incompatibilities in your pom.xml, and perform good pom.xml hygiene at all times. Make sure your infrastructure components such as Java, Maven, Python, npm, Node, and the rest are up-to-date and compatible. Please note: most of the examples in this book use Java 8 (and some examples rely on the advanced features present in Java 8), as well as using Maven 3+. Use java – version and mvn –version when in doubt!

6. Perform "overall optimization" throughout your technology stack. This includes at the Hadoop, Solr, and data source/sink levels. Identify bottlenecks and resource problems. Identify "problem hardware," particularly individual "problem processors," if you are running on a small Hadoop cluster.

7. Exercise the multicore functionality in your application frequently. It is rare you will use a single core in a sophisticated application, so make sure using more than one core works smoothly.

8. Perform integration testing religiously.

9. Performance monitoring is a must. Use a standard performance monitoring "script" and evaluate performance based on previous results as well as current expectations. Upgrade hardware and software as required to improve performance results, and re-monitor to insure accurate profiling.

10. Do not neglect unit tests. A good introduction to writing unit tests for current versions of Hadoop can be found at `https://wiki.apache.org/hadoop/HowToDevelopUnitTests`.

Apache Katta (http://katta.sourceforge.net/about) is a useful addition to any Solr-based distributed data pipelining architecture, and allows Hadoop indexing into shards, as well as many other advanced features.

HOW TO INSTALL AND CONFIGURE APACHE KATTA

1. Download Apache Katta from the repository at https://sourceforge.net/projects/katta/files/. Unzip the file.

2. Add the Katta environment variables to your .bash_profile file if you are running under MacOS, or the appropriate start-up file if running another version of Linux. These variables include (please note these are examples only; substitute your own appropriate path values here):

   ```
   export KATTA_HOME= /Users/kerryk/Downloads/kata-core-0.6.4
   ```

   and add the binary of Katta to the PATH so you can call it directly:

   ```
   export PATH=$KATTA_HOME/bin:$PATH
   ```

3. Check to make sure the Katta process is running correctly by typing

   ```
   ps -al | grep katta
   ```

   on the command line. You should see an output similar to Figure 6-4.

Figure 6-4. *A successful initialization of the Katta Solr subsystem*

4.   Successfully running the Katta component will produce results similar to those in Figure 6-4.

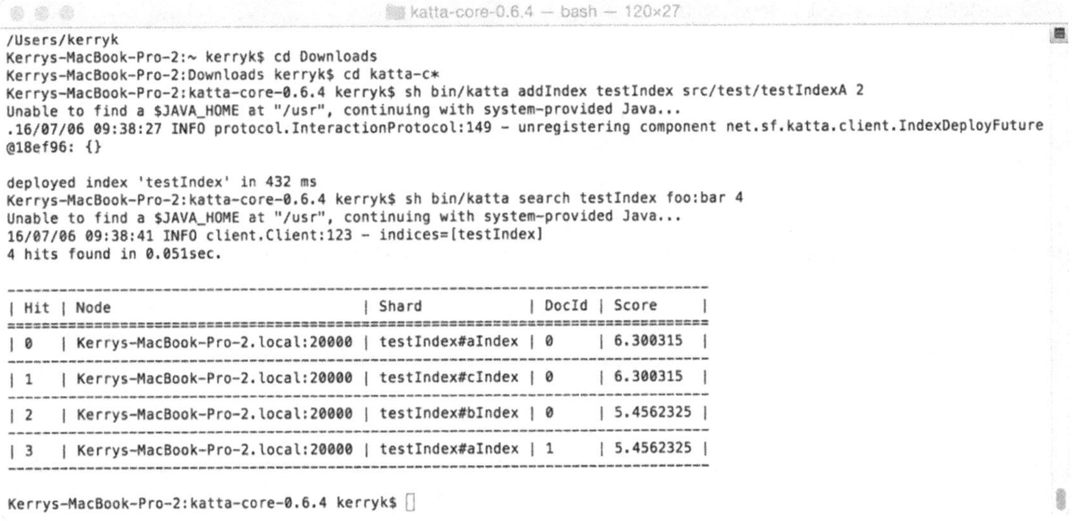

```
                                    katta-core-0.6.4 — bash — 120×27
/Users/kerryk
Kerrys-MacBook-Pro-2:~ kerryk$ cd Downloads
Kerrys-MacBook-Pro-2:Downloads kerryk$ cd katta-c*
Kerrys-MacBook-Pro-2:katta-core-0.6.4 kerryk$ sh bin/katta addIndex testIndex src/test/testIndexA 2
Unable to find a $JAVA_HOME at "/usr", continuing with system-provided Java...
.16/07/06 09:38:27 INFO protocol.InteractionProtocol:149 - unregistering component net.sf.katta.client.IndexDeployFuture
@18ef96: {}

deployed index 'testIndex' in 432 ms
Kerrys-MacBook-Pro-2:katta-core-0.6.4 kerryk$ sh bin/katta search testIndex foo:bar 4
Unable to find a $JAVA_HOME at "/usr", continuing with system-provided Java...
16/07/06 09:38:41 INFO client.Client:123 - indices=[testIndex]
4 hits found in 0.051sec.

----------------------------------------------------------------------------------
| Hit | Node                            | Shard           | DocId | Score     |
==================================================================================
| 0   | Kerrys-MacBook-Pro-2.local:20000 | testIndex#aIndex | 0     | 6.300315  |
----------------------------------------------------------------------------------
| 1   | Kerrys-MacBook-Pro-2.local:20000 | testIndex#cIndex | 0     | 6.300315  |
----------------------------------------------------------------------------------
| 2   | Kerrys-MacBook-Pro-2.local:20000 | testIndex#bIndex | 0     | 5.4562325 |
----------------------------------------------------------------------------------
| 3   | Kerrys-MacBook-Pro-2.local:20000 | testIndex#aIndex | 1     | 5.4562325 |
----------------------------------------------------------------------------------

Kerrys-MacBook-Pro-2:katta-core-0.6.4 kerryk$ []
```

Figure 6-5. *Successful installation and run of Apache Katta screen*

6.3   A Programming Example using SOLR

We are going to work through a complete example of using SOLR to load, modify, evaluate, and search a standard data set that we download from the Internet. We're going to highlight a few features of Solr as we go. As we noted earlier, Solr contains separate data repositories called "cores." Each one may have a separate defined schema associated with it. Solr cores may be created on the command line.

First, download the sample data set as a csv file from the URL http://samplecsvs.s3.amazonaws.com/SacramentocrimeJanuary2006.csv

You will find it in your downloads folder with the file name

yourDownLoadDirectory/SacramentocrimeJanuary2006.csv

Create a new SOLR core with the command:

./solr create -c crimecore1 -d basic_configs

You will see a screen similar to the one in Figure 6-2 if your core creation is successful.

```
                          ▨ bin — bash — 80×24
Last login: Thu Feb 18 16:26:46 on ttys009                           ▨
Now using node v0.12.0 (npm v2.5.1)
Kerrys-MBP:~ kerryk$ cd Downloads
Kerrys-MBP:Downloads kerryk$ cd solr-5*
Kerrys-MBP:solr-5.3.0 kerryk$ cd bin
Kerrys-MBP:bin kerryk$ ./solr create -c crimecore1 -d basic_configs
Unable to find a $JAVA_HOME at "/usr", continuing with system-provided Java...

Setup new core instance directory:
/Users/kerryk/Downloads/solr-5.3.0/server/solr/crimecore1

Creating new core 'crimecore1' using command:
https://localhost:8983/solr/admin/cores?action=CREATE&name=crimecore1&instanceDi
r=crimecore1

{
  "responseHeader":{
    "status":0,
    "QTime":63},
  "core":"crimecore1"}


Kerrys-MBP:bin kerryk$ ▯
```

Figure 6-6. *Successful construction of a Solr core*

Modify the schema file schema.xml by adding the right fields to the end of the specification.

```
<!- much more of the schema.xml file will be here -->

                       . . . . . . . . . . .

<!--   you will now add the field specifications for the cdatetime,address,district,beat,gri
d,crimedescr,ucr_ncic_code,latitude,longitude
   fields found in the data file SacramentocrimeJanuary2006.csv
-->
    <field name="cdatetime" type="string" indexed="true" stored="true" required="true"
    multiValued="false" />
    <field name="address" type="string" indexed="true" stored="true" required="true"
    multiValued="false" />

    <field name="district" type="string" indexed="true" stored="true" required="true"
    multiValued="false" />
<field name="beat" type="string" indexed="true" stored="true" required="true"
multiValued="false" />

<field name="grid" type="string" indexed="true" stored="true" required="true"
multiValued="false" />
<field name="crimedescr" type="string" indexed="true" stored="true" required="true"
multiValued="false" />
```

```
<field name="ucr_ncic_code" type="string" indexed="true" stored="true" required="true"
multiValued="false" />
<field name="latitude" type="string" indexed="true" stored="true" required="true"
multiValued="false" />

<field name="longitude" type="string" indexed="true" stored="true" required="true"
multiValued="false" />
    <field name="internalCreatedDate" type="date" indexed="true" stored="true"
required="true" multiValued="false" />

    <!-- the previous fields were added to the schema.xml file. Field type definition for
currentcy is shown below  -->

    <fieldType name="currency" class="solr.CurrencyField" precisionStep="8"
defaultCurrency="USD" currencyConfig="currency.xml" />

</schema>
```

It's easy to modify data by appending keys and additional data to the individual data lines of the CSV file. Listing 6-1 is a simple example of such a CSV conversion program.

Modify the Solr data by adding a unique key and creation date to the CSV file.

The program to do this is shown in Listing 6-1. The file name will be com/apress/converter/csv/CSVConverter.java.

The program to add fields to the CSV data set needs little explanation. It reads an input CSV file line by line, adding a unique ID and date field to each line of data. There are two helper methods within the class, createInternalSolrDate() and getCSVField().

Within the CSV data file, the header and the first few rows appear as in Figure 6-7, as shown in Excel.

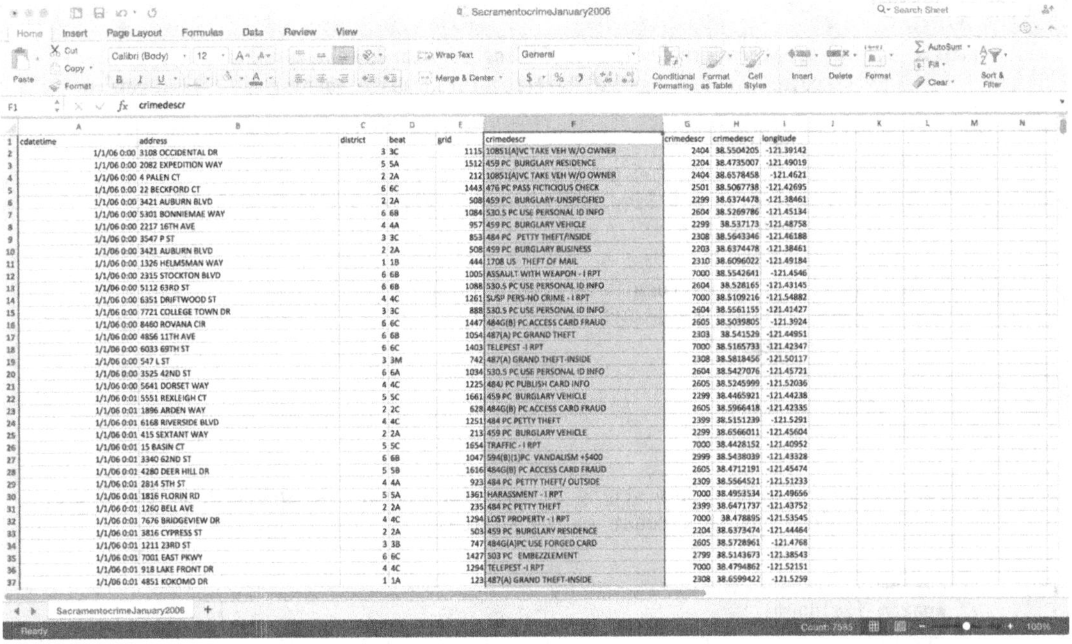

Figure 6-7. *Crime data CSV file. This data will be used throughout this chapter.*

Listing 6-1. Java source code for CSVConverter.java.

```java
package com.apress.converter.csv;

import java.io.File;
import java.io.FileNotFoundException;
import java.io.FileOutputStream;
import java.io.FileReader;
import java.io.FileWriter;
import java.io.IOException;
import java.io.LineNumberReader;
import java.text.DateFormat;
import java.text.SimpleDateFormat;
import java.util.ArrayList;
import java.util.Date;
import java.util.List;
import java.util.TimeZone;
import java.util.logging.Logger;

public class CSVConverter {
        Logger LOGGER = Logger.getAnonymousLogger();

        String targetSource = "SacramentocrimeJan2006.csv";
        String targetDest = "crime0.csv";

        /** Make a date Solr can understand from a regular oracle-style day string.
         *
         * @param regularJavaDate
         * @return
         */
        public String createInternalSolrDate(String regularJavaDate){
                if (regularJavaDate.equals("")||(regularJavaDate.equals("\"\""))){ return ""; }

                String answer = "";
                TimeZone tz = TimeZone.getTimeZone("UTC");
                DateFormat df = new SimpleDateFormat("yyyy-MM-dd'T'HH:mm'Z'");
                df.setTimeZone(tz);
                try {
                answer = df.format(new Date(regularJavaDate));
                } catch (IllegalArgumentException e){
                        return "";
                }
                return answer;
        }

        /** Get a CSV field in a CSV string by numerical index. Doesnt care if there are
            blank fields, but they count in the indices.
         *
         * @param s
         * @param fieldnum
         * @return
         */
```

```java
public String getCSVField(String s, int fieldnum){
        String answer = "";
        if (s != null) { s = s.replace(",,", " ", ,",");
        String[] them = s.split(",");
        int count = 0;
        for (String t : them){
                if (fieldnum == count) answer = them[fieldnum];
                count++;
        }
        }
        return answer;
}


public CSVConverter(){
        LOGGER.info("Performing CSV conversion for SOLR input");

        List<String>contents = new ArrayList<String>();
        ArrayList<String>result = new ArrayList<String>();
        String readline = "";
        LineNumberReader reader = null;
        FileOutputStream writer = null;
        try {
                reader = new LineNumberReader(new FileReader(targetSource));
                writer = new FileOutputStream (new File(targetDest));
                int count = 0;
                int thefield = 1;
                while (readline != null){
                readline = reader.readLine();
                if (readline.split(","))<2){
                        LOGGER.info("Last line, exiting...");
                        break;
                }
                if (count != 0){
                        String origDate = getCSVField(readline, thefield).split(" ")[0];
                        String newdate = createInternalSolrDate(origDate);
                        String resultLine = readline + "," + newdate+"\n";
                        LOGGER.info("===== Created new line: " + resultLine);
                        writer.write(resultLine.getBytes());
                        result.add(resultLine);
                } else {
                        String resultLine = readline +",INTERNAL_CREATED_DATE\n";
                          // add the internal date for faceted search
                        writer.write(resultLine.getBytes());
                }
                count++;
                LOGGER.info("Just read imported row: " + readline);
                }
        } catch (FileNotFoundException e) {
                e.printStackTrace();
        } catch (IOException e) {
                // TODO Auto-generated catch block
```

101

```
                    e.printStackTrace();
            }
            for (String line : contents){
            String newLine = "";
            }
            try {
            reader.close();
            writer.close();
            } catch (IOException e){ e.printStackTrace(); }
            LOGGER.info("...CSV conversion complete...");
      }

      /** MAIN ROUTINE
       *
       * @param args
       */
      public static void main(String[] args){
            new CSVConverter(args[0], args[1]);
      }

}
```

Compile the file by typing:

```
javac com/apress/converter/csv/CSVConverter.java
```

After setting up the CSV conversion program properly as described above, you can run it by typing

```
java com.apress.converter.csv.CSVConverter inputcsvfile.csv outputcsvfile.csv
```

Post the modified data to the SOLR core:

```
./post -c crimecore1 ./modifiedcrimedata2006.csv
```

Now that we've posted the data to the Solr core, we can examine the data set in the Splr dashboard. Go to localhost:8983 to do this. You should see a screen similar to the one in Figure 6-4.

Figure 6-8. *Initial Solr dashboard*

```
{
  "responseHeader":{
    "status":0,
    "QTime":1,
    "params":{
      "q":"title:*MONEY*",
      "indent":"true",
      "wt":"json"}},
  "response":{"numFound":250,"start":0,"docs":[
    {
      "id":"3001",
      "title":["U.K. MONEY MARKET SHORTAGE FORECAST AT 250 MLN STG"],
      "dateline":["LONDON, March 9 -"],
      "text":["The Bank of England said it forecast a\nshortage of around 250 mln stg in the money market today.\n    Among the
factors affecting liquidity, it said bills\nmaturing in official hands and the treasury bill take-up would\ndrain around 1.02
billion stg while below target bankers'\nbalances would take out a further 140 mln.\n    Against this, a fall in the note
circulation would add 345\nmln stg and the net effect of exchequer transactions would be\nan inflow of some 545 mln stg, the Bank
added.\n REUTER"],
      "places":["uk"],
      "topics":["money-fx"],
      "date":["1987-03-09T12:58:41.012Z"],
      "_version_":1516128122063290368},
    {
      "id":"3002",
      "title":["BANK OF FRANCE SETS MONEY MARKET TENDER"],
      "dateline":["PARIS, March 9 -"],
      "text":["The Bank of France said it invited offers\nof first category paper today for a money market
intervention\ntender.\n    Money market dealers said conditions seemed right for the\nBank to cut its intervention rate at the
tender by a quarter\npercentage point to 7-3/4 pct from eight, reflecting an easing\nin call money rate last week, and the French
franc's steadiness\non foreign exchange markets since the February 22 currency\nstabilisation accord here by the Group of Five and
Canada.\n    Intervention rate was last raised to eight pct from 7-1/4\non January 2. Call money today was quoted at 7-11/16 7-3/4
pct.\n REUTER"],
      "places":["france"],
      "topics":["money-fx",
        "interest"],
      "date":["1987-03-09T13:03:09.075Z"],
      "_version_":1516128122066436096},
    {
      "id":"3044",
      "title":["U.K. MONEY MARKET GIVEN FURTHER HELP AT NEW RATES"],
      "dateline":["LONDON, March 9 -"],
      "text":["The Bank of England said it provided the\nmarket with further assistance during the afternoon, buying\nbills
worth 166 mln stg at the lower rates introduced this\nmorning.\n    It bought 45 mln stg of local authority bills plus 27 mln\nstg
of bank bills in band one at 10-3/8 pct together with 94\nmln stg of band two bank bills at 10-5/16 pct.\n    The bank also
revised its estimate of the market shortage\nup to 300 mln stg from 250 mln this morning. It has given total\nassistance of 213
mln stg today.\n REUTER"],
      "places":["uk"],
      "topics":["money-fx",
        "interest"],
      "date":["1987-03-09T17:30:35.065Z"],
      "_version_":1516128122220576768},
    {
```

Figure 6-9. *Result of Solr query, showing the JSON output format*

We can also evaluate data from the Solandra core we created earlier in the chapter, as shown in Figure _ _.

Now select the crimedata0 core from the Core Selector drop-down. Click on query and change the output format ('wt' parameter dropdown) to csv , so that you can see several lines of data at once. You will see a data display similar to the one in Figure 6-9.

Figure 6-10. *Result of Solr query using the dashboard (Sacramento crime data core)*

Because of Solr's RESTful interface, we can make queries either through the dashboard (conforming to Lucene's query syntax discussed earlier) or on the command line using the CURL utility.

6.4 Using the ELK Stack (Elasticsearch, Logstash, and Kibana)

As we mentioned before, there are alternatives to Lucene, Solr, and Nutch. Depending on the overall architecture of your system, a variety of technology stacks, languages, integration and plug-in helper libraries, and functionality are available to you. Some of these components may use Lucene or Solr, or be compatible with Lucene/Solr components through integration helper libraries, such as Spring Data, Spring MVC, or Apache Camel, among others. An example of an alternative to the basic Lucene stack, known as the "ELK stack," is shown in Figure 6-6.

Elasticsearch (elasticsearch.org) is a distributed high-performance search engine . Under the hood, Elasticsearch uses Lucene as a core component, as shown in Figure 6-3. Elasticsearch is a strong competitor to SolrCloud, and is easy to scale out, maintain, monitor, and deploy.

Why would you use Elasticsearch instead of Solr? Taking a careful look at the feature matrices for Solr and Elasticsearch reveals that, in many ways, the two toolkits have similar functionality. They both leverage Apache Lucene. Both Solr and Elasticsearch can use JSON as a data exchange format, although Solr also supports XML.

Table 6-2. *Feature comparison table of Elasticsearch features vs. Apache Solr features*

	JSON	XML	CSV	HTTP REST	JMX	Client Libraries	Lucene Query Parsing	Self Contained Distributed Cluster	Sharding	Visualization	Web Admin Interface
Solr	X	X	X	X	X	Java	X		X	Kibana Port (Banana)	
Elastic Search	X			X		Java Python Javascript	X	X	X	Kibana	

Logstash (logstash.net) is a useful application to allow importing of a variety of different kinds of data into Elasticsearch, including CSV-formatted files and ordinary "log format" files. Kibana (`https://www.elastic.co/guide/en/kibana/current/index.html`) is an open source visualization component which allows customizable . Together Elasticsearch, Logstash, and Kibana form the so-called "ELK stack," which can be principally used to. In this section, we'll look at a small example of the ELK stack in action.

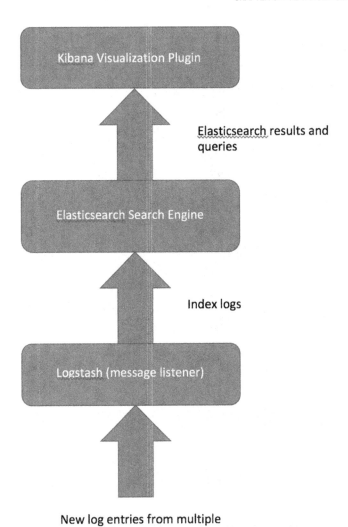

Figure 6-11. *The so-called "ELK stack": Elasticsearch, Logstash, and Kibana visualization*

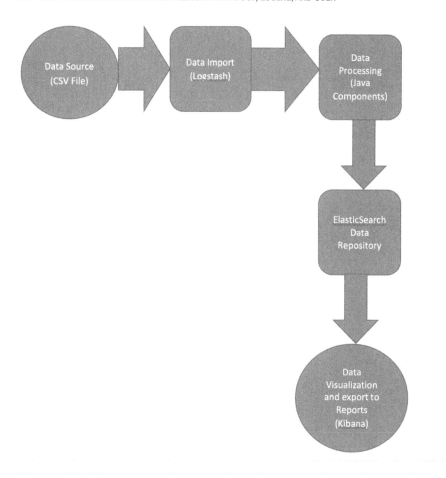

Figure 6-12. *ELK stack in use: Elasticsearch search engine/pipeline architecture diagram*

INSTALLING ELASTICSEARCH, LOGSTASH, AND KIBANA

Installing and trying out the ELK Stack couldn't be easier. It is a familiar process if you have followed through the introductory chapters of the book so far. Follow the three steps below to install and test the ELK stack:

1. Download Elasticsearch from `https://www.elastic.co/downloads/elasticsearch`.

   Unzip the downloaded file to a convenient staging area. Then,

   ```
   cd $ELASTICSEARCH_HOME/bin/
   ./elasticsearch
   ```

   Elasticsearch will start up successfully with a display similar to that in Figure 6-3.

```
Kerrys-MacBook-Pro:bin kerryk$ ./elasticsearch
Unable to find a $JAVA_HOME at "/usr", continuing with system-provided Java...
[2016-04-15 21:51:45,351][INFO ][node                     ] [Lady Lark] version[2.3.1], pid[35066], build[bd98092/2016-04-04T12:25:05Z]
[2016-04-15 21:51:45,352][INFO ][node                     ] [Lady Lark] initializing ...
[2016-04-15 21:51:45,733][INFO ][plugins                  ] [Lady Lark] modules [reindex, lang-expression, lang-groovy], plugins [], sites []
[2016-04-15 21:51:45,749][INFO ][env                      ] [Lady Lark] using [1] data paths, mounts [[/ (/dev/disk1)]], net usable_space [251.1
gb], net total_space [464.7gb], spins? [unknown], types [hfs]
[2016-04-15 21:51:45,751][INFO ][env                      ] [Lady Lark] heap size [989.8mb], compressed ordinary object pointers [true]
[2016-04-15 21:51:45,751][WARN ][env                      ] [Lady Lark] max file descriptors [10240] for elasticsearch process likely too low, c
onsider increasing to at least [65536]
[2016-04-15 21:51:46,867][INFO ][node                     ] [Lady Lark] initialized
[2016-04-15 21:51:46,867][INFO ][node                     ] [Lady Lark] starting ...
[2016-04-15 21:51:46,917][INFO ][transport                ] [Lady Lark] publish_address {127.0.0.1:9300}, bound_addresses {[fe80::1]:9300}, {[::
1]:9300}, {127.0.0.1:9300}
[2016-04-15 21:51:46,920][INFO ][discovery                ] [Lady Lark] elasticsearch/bNbMkEtIR-mBDBxA7QX8TA
[2016-04-15 21:51:49,945][INFO ][cluster.service          ] [Lady Lark] new_master {Lady Lark}{bNbMkEtIR-mBDBxA7QX8TA}{127.0.0.1}{127.0.0.1:9300
}, reason: zen-disco-join(elected_as_master, [0] joins received)
[2016-04-15 21:51:49,969][INFO ][http                     ] [Lady Lark] publish_address {127.0.0.1:9200}, bound_addresses {[fe80::1]:9200}, {[::
1]:9200}, {127.0.0.1:9200}
[2016-04-15 21:51:49,969][INFO ][node                     ] [Lady Lark] started
[2016-04-15 21:51:50,046][INFO ][gateway                  ] [Lady Lark] recovered [11] indices into cluster_state
[2016-04-15 21:51:50,951][INFO ][cluster.routing.allocation] [Lady Lark] Cluster health status changed from [RED] to [YELLOW] (reason: [shards s
tarted [[.kibana][0], [.kibana][0]] ...]).
```

Figure 6-13. *Successful start-up of the Elasticsearch server from the binary directory*

Use the following Java program to import the crime data CSV file (or, with a little modification, any CSV formatted data file you wish):

```java
public static void main(String[] args)
    {
        System.out.println( "Import crime data" );
        String originalClassPath = System.getProperty("java.class.path");
        String[] classPathEntries = originalClassPath.split(";");
        StringBuilder esClasspath = new StringBuilder();
        for (String entry : classPathEntries) {
        if (entry.contains("elasticsearch") || entry.contains("lucene")) {
        esClasspath.append(entry);
        esClasspath.append(";");
        }
        }
        System.setProperty("java.class.path", esClasspath.toString());
        System.setProperty("java.class.path", originalClassPath);
        System.setProperty("es.path.home", "/Users/kerryk/Downloads/elasticsearch-2.3.1");
        String file = "SacramentocrimeJanuary2006.csv";
        Client client = null;
        try {

        client = TransportClient.builder().build()
          .addTransportAddress(new InetSocketTransportAddress(InetAddress.getByName("localho
              st"), 9300));

        int numlines = 0;
        XContentBuilder builder = null;

        int i=0;

        String currentLine = "";
        BufferedReader br = new BufferedReader(new FileReader(file));
```

```java
    while ((currentLine = br.readLine()) != null) {
    if (i > 0){
    System.out.println("Processing line: " + currentLine);
String[] tokens = currentLine.split(",");
String city = "sacramento";
String recordmonthyear = "jan2006";
String cdatetime = tokens[0];
String address = tokens[1];
String district = tokens[2];
String beat = tokens[3];
String grid = tokens[4];
String crimedescr = tokens[5];
String ucrnciccode = tokens[6];
String latitude = tokens[7];
String longitude = tokens[8];
System.out.println("Crime description = " + crimedescr);
i=i+1;
System.out.println("Index is: " + i);
IndexResponse response = client.prepareIndex("thread", "answered", "400"+new
    Integer(i).toString()).setSource(

jsonBuilder()
.startObject()
.field("cdatetime", cdatetime)
.field("address", address)
.field("district", district)
.field("beat", beat)
.field("grid", grid)
.field("crimedescr", crimedescr)
.field("ucr_ncic_code", ucrnciccode)
.field("latitude", latitude)
.field("longitude", longitude)
.field("entrydate", new Date())
.endObject())
.execute().actionGet();

        } else {
            System.out.println("Ignoring first line...");
            i++;
        }
    }

} catch (Exception e) {
    // TODO Auto-generated catch block
    e.printStackTrace();

    }
    }
```

Run the program in Eclipse or in the command line. You will see a result similar to the one in Figure 6-14. Please note that each row of the CSV is entered as a set of fields into the Elasticsearch repository. You can also select the index name and index type by changing the appropriate strings in the code example.

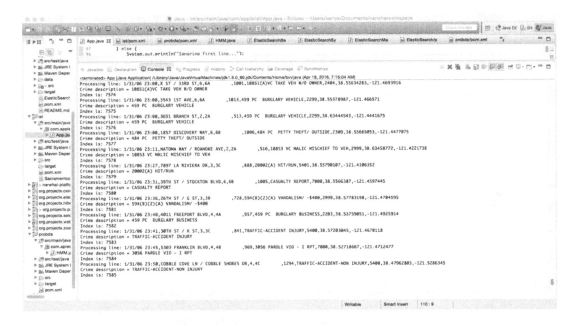

Figure 6-14. *Successful test of an Elasticsearch crime database import from the Eclipse IDE*

You can test the query capabilities of your new Elasticsearch set-up by using 'curl' on the command line to execute some sample queries, such as:

```
[2016-04-19 10:30:58,288][INFO ][discovery              ] [Redneck] elasticsearch/ZTCCH2BDRLaWeKyLv7XXeQ
[2016-04-19 10:31:01,321][INFO ][cluster.service       ] [Redneck] new_master {Redneck}{ZTCCH2BDRLaWeKyLv7XXeQ}{127.0.0.1}{127.0.0.1:9300}, r
eason: zen-disco-join(elected_as_master, [0] joins received)
[2016-04-19 10:31:01,354][INFO ][http                  ] [Redneck] publish_address {127.0.0.1:9200}, bound_addresses {[fe80::1]:9200}, {[::1]
:9200}, {127.0.0.1:9200}
[2016-04-19 10:31:01,354][INFO ][node                  ] [Redneck] started
[2016-04-19 10:31:01,483][INFO ][gateway               ] [Redneck] recovered [14] indices into cluster_state
[2016-04-19 10:31:02,535][INFO ][cluster.routing.allocation] [Redneck] Cluster health status changed from [RED] to [YELLOW] (reason: [shards sta
rted [[.kibana][0]] ...]).
[2016-04-19 10:33:32,555][INFO ][cluster.metadata      ] [Redneck] [crime2] update_mapping [crime2]
[2016-04-19 11:56:12,984][INFO ][cluster.metadata      ] [Redneck] [threads2] creating index, cause [api], templates [], shards [5]/[1], mapp
ings [messages2]
[2016-04-19 11:56:13,042][INFO ][cluster.routing.allocation] [Redneck] Cluster health status changed from [RED] to [YELLOW] (reason: [shards sta
rted [[threads2][4]] ...]).
[2016-04-19 11:56:57,926][INFO ][cluster.metadata      ] [Redneck] [threads2] update_mapping [messages2]
```

Figure 6-15. *You can see the schema update logged in the Elasticsearch console*

```
Kerrys-MBP:~ kerryk$ curl -XGET 'http://127.0.0.1:9200/thread/_search?pretty=true' -
d '
{
    "query" : {
        "matchAll" : {}
    }
}'
{
    "took" : 2,
    "timed_out" : false,
    "_shards" : {
        "total" : 5,
        "successful" : 5,
        "failed" : 0
    },
    "hits" : {
        "total" : 22773,
        "max_score" : 1.0,
        "hits" : [ {
            "_index" : "thread",
            "_type" : "answered",
            "_id" : "2004150",
            "_score" : 1.0,
            "_source" : {
                "cdatetime" : "1/18/06 9:00",
                "district" : "2820 CONNIE DR",
                "beat" : "2",
                "grid" : "2C      ",
                "crimedescr" : "567",
                "ucr_ncic_code" : "484G(B) PC ACCESS CARD FRAUD",
                "latitude" : "2605",
                "longitude" : "38.61726903",
                "entrydate" : "2016-04-14T23:07:07.292Z"
            }
        }, {
            "_index" : "thread",
            "_type" : "answered",
            "_id" : "2004153",
            "_score" : 1.0,
            "_source" : {
                "cdatetime" : "1/18/06 9:16",
                "district" : "914 12TH ST",
                "beat" : "3",
                "grid" : "3M      ",
                "crimedescr" : "734",
                "ucr_ncic_code" : "290(A)(1)(D) FAIL TO REG-5 DAY",
                "latitude" : "3699",
                "longitude" : "38.58029569",
                "entrydate" : "2016-04-14T23:07:07.299Z"
            }
        }, {
            "_index" : "thread",
            "_type" : "answered",
            "_id" : "2004162",
            "_score" : 1.0,
            "_source" : {
                "cdatetime" : "1/18/06 10:00",
```

Figure 6-16. Successful test of an Elasticsearch crime database query from the command line

2. Download Logstash from `https://www.elastic.co/downloads/logstash`. Unzip
    the downloaded file to the staging area.

    ```
    cd <your logstash staging area, LOGSTASH_HOME>
    ```

    After entering some text, you will see an echoed result similar to Figure 6-6.

    You will also need to set up a configuration file for use with Logstash. Follow the
    directions found at to make a configuration file such as the one shown in Listing 6-2.

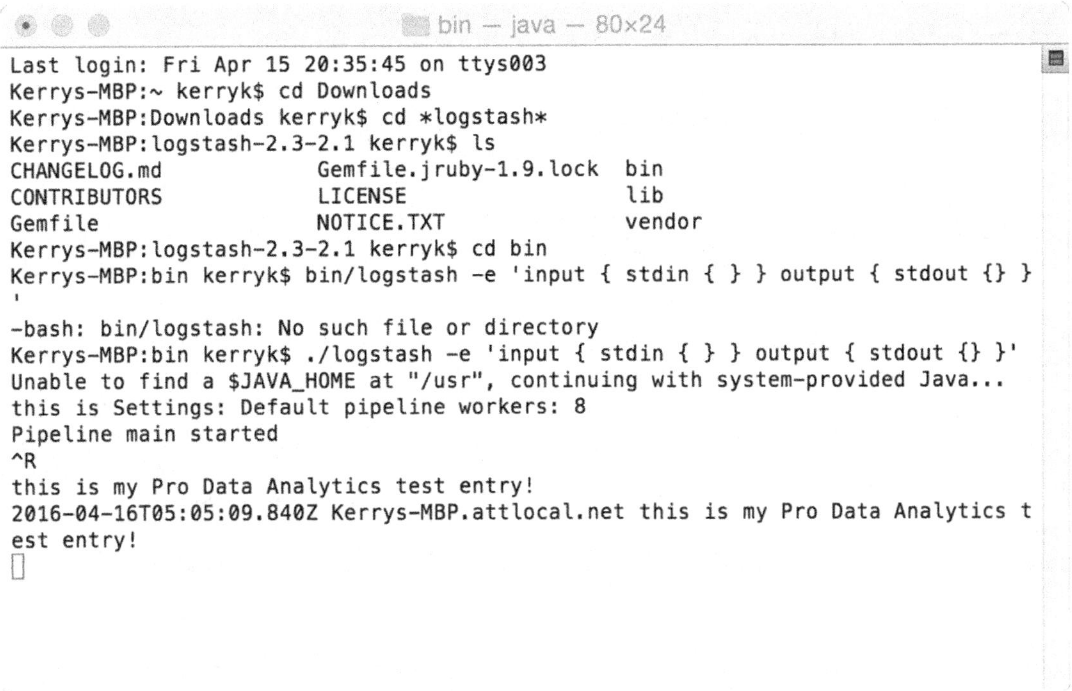

```
                                          bin — java — 80×24
Last login: Fri Apr 15 20:35:45 on ttys003
Kerrys-MBP:~ kerryk$ cd Downloads
Kerrys-MBP:Downloads kerryk$ cd *logstash*
Kerrys-MBP:logstash-2.3-2.1 kerryk$ ls
CHANGELOG.md              Gemfile.jruby-1.9.lock  bin
CONTRIBUTORS              LICENSE                 lib
Gemfile                   NOTICE.TXT              vendor
Kerrys-MBP:logstash-2.3-2.1 kerryk$ cd bin
Kerrys-MBP:bin kerryk$ bin/logstash -e 'input { stdin { } } output { stdout {} }
'
-bash: bin/logstash: No such file or directory
Kerrys-MBP:bin kerryk$ ./logstash -e 'input { stdin { } } output { stdout {} }'
Unable to find a $JAVA_HOME at "/usr", continuing with system-provided Java...
this is Settings: Default pipeline workers: 8
Pipeline main started
^R
this is my Pro Data Analytics test entry!
2016-04-16T05:05:09.840Z Kerrys-MBP.attlocal.net this is my Pro Data Analytics t
est entry!
```

Figure 6-17. *Testing your Logstash installation. You can enter text from the command line.*

Listing 6-2. Typical Logstash configuration file listing

```
input { stdin { } }

filter {
  grok {
    match => { "message" => "%{COMBINEDAPACHELOG}" }
  }
  date {
    match => [ "timestamp" , "dd/MMM/yyyy:HH:mm:ss Z" ]
  }
}

output {
  elasticsearch { hosts => ["localhost:9200"] }
  stdout { codec => rubydebug }
}
```

3. Download Kibana from https://www.elastic.co/downloads/kibana.

   Unzip the downloaded file to the staging area.

   In a similar way to starting the Elasticsearch server:

   ```
   cd bin
   ./kibana
   ```

113

```
●  ⊛  ⊛                                    ▤ bin — node — 117×49
Kerrys-MacBook-Pro:kibana-4.5.0-darwin-x64 2 kerryk$ cd bin                                              ▤
Kerrys-MacBook-Pro:bin kerryk$ ./kibana
  log   [11:43:48.874] [info][status][plugin:kibana] Status changed from uninitialized to green – Ready
  log   [11:43:49.126] [info][status][plugin:elasticsearch] Status changed from uninitialized to yellow – Waiting for
Elasticsearch
  log   [11:43:49.257] [info][status][plugin:kbn_vislib_vis_types] Status changed from uninitialized to green – Ready
  log   [11:43:49.292] [info][status][plugin:markdown_vis] Status changed from uninitialized to green – Ready
  log   [11:43:49.377] [info][status][plugin:metric_vis] Status changed from uninitialized to green – Ready
  log   [11:43:49.428] [info][status][plugin:spyModes] Status changed from uninitialized to green – Ready
  log   [11:43:49.461] [info][status][plugin:statusPage] Status changed from uninitialized to green – Ready
  log   [11:43:49.560] [info][status][plugin:table_vis] Status changed from uninitialized to green – Ready
  log   [11:43:49.611] [info][listening] Server running at http://0.0.0.0:5601
  log   [11:43:49.629] [info][status][plugin:elasticsearch] Status changed from yellow to green – Kibana index ready
  log   [15:45:20.158] [error][status][plugin:elasticsearch] Status changed from green to red – This version of Kiban
a requires Elasticsearch ^2.3.0 on all nodes. I found the following incompatible nodes in your cluster: Elasticsearch
 v1.5.2 @ inet[/17.115.177.187:9200] (17.115.177.187)
  log   [15:45:32.751] [info][status][plugin:elasticsearch] Status changed from red to green – Kibana index ready
  log   [15:56:13.491] [error][status][plugin:elasticsearch] Status changed from green to red – This version of Kiban
a requires Elasticsearch ^2.3.0 on all nodes. I found the following incompatible nodes in your cluster: Elasticsearch
 v1.5.2 @ inet[/17.115.177.187:9200] (17.115.177.187)
  log   [15:56:46.162] [error][status][plugin:elasticsearch] Status changed from red to red – Elasticsearch is still
initializing the kibana index.
  log   [15:56:48.674] [info][status][plugin:elasticsearch] Status changed from red to green – Kibana index ready
  log   [16:06:16.594] [error][status][plugin:elasticsearch] Status changed from green to red – This version of Kiban
a requires Elasticsearch ^2.3.0 on all nodes. I found the following incompatible nodes in your cluster: Elasticsearch
 v1.5.2 @ inet[/17.115.177.187:9200] (17.115.177.187)
  log   [16:06:22.037] [error][status][plugin:elasticsearch] Status changed from red to red – Elasticsearch is still
initializing the kibana index.
  log   [16:06:24.546] [info][status][plugin:elasticsearch] Status changed from red to green – Kibana index ready
  log   [16:39:20.422] [error][status][plugin:elasticsearch] Status changed from green to red – Request Timeout after
3000ms
  log   [18:50:40.150] [info][status][plugin:elasticsearch] Status changed from red to green – Kibana index ready
  log   [19:09:06.389] [error][status][plugin:elasticsearch] Status changed from green to red – Request Timeout after
3000ms
  log   [19:09:08.902] [info][status][plugin:elasticsearch] Status changed from red to green – Kibana index ready
  log   [19:40:59.476] [error][status][plugin:elasticsearch] Status changed from green to red – Request Timeout after
3000ms
  log   [19:41:08.800] [info][status][plugin:elasticsearch] Status changed from red to green – Kibana index ready
  log   [20:05:26.004] [error][status][plugin:elasticsearch] Status changed from green to red – Request Timeout after
3000ms
  log   [20:05:31.425] [info][status][plugin:elasticsearch] Status changed from red to green – Kibana index ready
  log   [20:17:24.683] [error][status][plugin:elasticsearch] Status changed from green to red – Request Timeout after
3000ms
  log   [20:17:27.193] [info][status][plugin:elasticsearch] Status changed from red to green – Kibana index ready
  log   [21:40:30.077] [error][status][plugin:elasticsearch] Status changed from green to red – Request Timeout after
3000ms
  log   [21:40:34.103] [info][status][plugin:elasticsearch] Status changed from red to green – Kibana index ready
  log   [21:51:43.581] [error][status][plugin:elasticsearch] Status changed from green to red – This version of Kiban
a requires Elasticsearch ^2.3.0 on all nodes. I found the following incompatible nodes in your cluster: Elasticsearch
```

Figure 6-18. *Successful start-up of the Kibana visualization component from its binary directory*

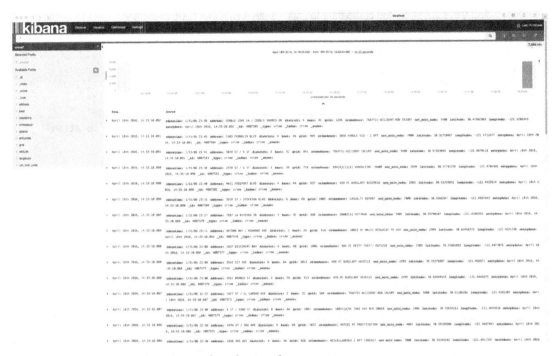

Figure 6-19. *Kibana dashboard example with crime dataset*

You can easily query for keywords or more complex queries interactively using the Kibana dashboard as shown in Figure 6-19.

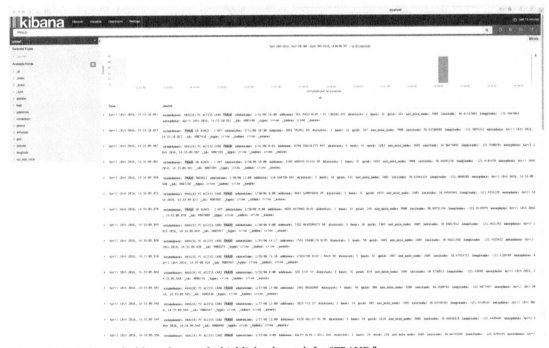

Figure 6-20. *Kibana dashboard example: highlighted search for "FRAUD."*

115

Add this schema for the crime data to Elasticsearch with this cURL command:

```
curl -XPUT http://localhost:9200/crime2 -d '
{ "mappings" :
{ "crime2" : { "properties" : { "cdatetime" : {"type" : "string"}, "address" : {"type":
"string"}, "district" : {"type" : "string"}, "beat": {"type" : "string"}, "grid":
{"type" : "string"}, "crimedescr": {"type": "string"}, "ucr_ncic_code": {"type":
"string"},"latitude": {"type" : "string"}, "longitude": {"type" : "string"}, "location":
{"type" : "geo_point"}}
} } }'
```

Notice the "location" tag in particular, which has a geo_point-type definition. This allows Kibana to identify the physical location on a map for visualization purposes, as shown in Figure 6-21.

Figure 6-21. *The crime data for Sacramento as a visualization in Kibana*

Figure 6-21 is a good example of understanding a complex data set at a glance. We can immediately pick out the "high crime" areas in red.

6.5 Solr vs. ElasticSearch : Features and Logistics

In this section we will use as an example, the so-called CRUD operations (create, replace, update, and delete methods, with an additional search utility method) in a code example using Elasticsearch.

Listing 6-3. CRUD operations for Elasticsearch example

```java
package com.apress.main;

import java.io.IOException;
import java.util.Date;
import java.util.HashMap;
import java.util.Map;
import org.elasticsearch.action.delete.DeleteResponse;
import org.elasticsearch.action.get.GetResponse;
import org.elasticsearch.action.search.SearchResponse;
import org.elasticsearch.action.search.SearchType;
import org.elasticsearch.client.Client;
import static org.elasticsearch.index.query.QueryBuilders.fieldQuery;
import org.elasticsearch.node.Node;
import static org.elasticsearch.node.NodeBuilder.nodeBuilder;
import org.elasticsearch.search.SearchHit;

/**
 *
 * @author kerryk
 */

public class ElasticSearchMain {

        public static final String INDEX_NAME = "narwhal";
        public static final String THEME_NAME = "messages";

    public static void main(String args[]) throws IOException{

        Node node    = nodeBuilder().node();
        Client client = node.client();

        client.prepareIndex(INDEX_NAME, THEME_NAME, "1")
            .setSource(put("ElasticSearch: Java",
                        "ElasticSeach provides Java API, thus it executes
                          all operations " +
                         "asynchronously by using client object..",
                        new Date(),
                        new String[]{"elasticsearch"},
                        "Kerry Koitzsch", "iPad", "Root")).execute().actionGet();
```

```
        client.prepareIndex(INDEX_NAME, THEME_NAME, "2")
                .setSource(put("Java Web Application and ElasticSearch (Video)",
                                        "Today, here I am for exemplifying the usage of
ElasticSearch which is an open source, distributed " +
                                        "and scalable full text search engine and a data
analysis tool in a Java web application.",
                                        new Date(),
                                        new String[]{"elasticsearch"},
                                        "Kerry Koitzsch", "Apple TV", "Root")).execute().
                                            actionGet();


        get(client, INDEX_NAME, THEME_NAME, "1");

        update(client, INDEX_NAME, THEME_NAME, "1", "title", "ElasticSearch: Java API");
        update(client, INDEX_NAME, THEME_NAME, "1", "tags", new String[]{"bigdata"});

        get(client, INDEX_NAME, THEME_NAME, "1");

        search(client, INDEX_NAME, THEME_NAME, "title", "ElasticSearch");

        delete(client, INDEX_NAME, THEME_NAME, "1");

        node.close();
    }

    public static Map<String, Object> put(String title, String content, Date postDate,
                                            String[] tags, String author,
                                            String communityName, String
                                            parentCommunityName){

        Map<String, Object> jsonDocument = new HashMap<String, Object>();

        jsonDocument.put("title", title);
        jsonDocument.put("content", content);
        jsonDocument.put("postDate", postDate);
        jsonDocument.put("tags", tags);
        jsonDocument.put("author", author);
        jsonDocument.put("communityName", communityName);
        jsonDocument.put("parentCommunityName", parentCommunityName);
        return jsonDocument;
    }

    public static void get(Client client, String index, String type, String id){

        GetResponse getResponse = client.prepareGet(index, type, id)
                                        .execute()
                                        .actionGet();
        Map<String, Object> source = getResponse.getSource();

        System.out.println("------------------------------");
        System.out.println("Index: " + getResponse.getIndex());
```

```java
        System.out.println("Type: " + getResponse.getType());
        System.out.println("Id: " + getResponse.getId());
        System.out.println("Version: " + getResponse.getVersion());
        System.out.println(source);
        System.out.println("----------------------------");
    }

    public static void update(Client client, String index, String type,
                                    String id, String field, String newValue){

        Map<String, Object> updateObject = new HashMap<String, Object>();
        updateObject.put(field, newValue);

        client.prepareUpdate(index, type, id)
            .setScript("ctx._source." + field + "=" + field)
            .setScriptParams(updateObject).execute().actionGet();
    }

    public static void update(Client client, String index, String type,
                                    String id, String field, String[] newValue){

        String tags = "";
        for(String tag :newValue)
            tags += tag + ", ";

        tags = tags.substring(0, tags.length() - 2);

        Map<String, Object> updateObject = new HashMap<String, Object>();
        updateObject.put(field, tags);

        client.prepareUpdate(index, type, id)
            .setScript("ctx._source." + field + "+=" + field)
            .setScriptParams(updateObject).execute().actionGet();
    }

    public static void search(Client client, String index, String type,
                                    String field, String value){

        SearchResponse response = client.prepareSearch(index)
                                .setTypes(type)
                                .setSearchType(SearchType.QUERY_AND_FETCH)
                                .setQuery(fieldQuery(field, value))
                                .setFrom(0).setSize(60).setExplain(true)
                                .execute()
                                .actionGet();

        SearchHit[] results = response.getHits().getHits();

        System.out.println("Current results: " + results.length);
        for (SearchHit hit : results) {
```

```
            System.out.println("-----------------------------");
            Map<String,Object> result = hit.getSource();
            System.out.println(result);
        }
    }

    public static void delete(Client client, String index, String type, String id){

        DeleteResponse response = client.prepareDelete(index, type, id).execute().actionGet();
        System.out.println("===== Information on the deleted document:");
        System.out.println("Index: " + response.getIndex());
        System.out.println("Type: " + response.getType());
        System.out.println("Id: " + response.getId());
        System.out.println("Version: " + response.getVersion());
    }
}
```

Defining the CRUD operations for a search component is key to the overall architecture and logistics of how the customized component will "fit in" with the rest of the system.

6.6   Spring Data Components with Elasticsearch and Solr

In this section, we will develop a code example which uses Spring Data to implement the same kind of component using Solr and Elasticsearch as the search frameworks being used "under the hood."

You can define the two properties for Elasticsearch and Solr respectively for your pom.xml file as shown here:

```
<spring.data.elasticsearch.version>2.0.1.RELEASE</spring.data.elasticsearch.version>
<spring.data.solr.version>2.0.1.RELEASE</spring.data.solr.version>

<dependency>
        <groupId>org.springframework.data</groupId>
        <artifactId>spring-data-elasticsearch</artifactId>
        <version>2.0.1.RELEASE</version>
</dependency>
and
<dependency>
        <groupId>org.springframework.data</groupId>
        <artifactId>spring-data-solr</artifactId>
        <version>2.0.1.RELEASE</version>
</dependency>
```

We can now develop Spring Data-based code examples as shown in Listing 6-5 and Listing 6-6.

Listing 6-4. NLP program—main() executable method

```
package com.apress.probda.solr.search;
import org.springframework.boot.SpringApplication;
import org.springframework.boot.autoconfigure.EnableAutoConfiguration;
import org.springframework.context.annotation.ComponentScan;
import org.springframework.context.annotation.Configuration;
import org.springframework.context.annotation.Import;
import com.apress.probda.context.config.SearchContext;
import com.apress.probda.context.config.WebContext;
@Configuration
@ComponentScan
@EnableAutoConfiguration
@Import({ WebContext.class, SearchContext.class })
public class Application {
        public static void main(String[] args) {
                SpringApplication.run(Application.class, args);
        }
import org.apache.solr.client.solrj.SolrServer;
import org.apache.solr.client.solrj.impl.HttpSolrServer;
import org.springframework.beans.factory.annotation.Value;
import org.springframework.context.annotation.Bean;
import org.springframework.context.annotation.Configuration;
import org.springframework.data.solr.repository.config.EnableSolrRepositories;


@Configuration
@EnableSolrRepositories(basePackages = { "org.springframework.data.solr.showcase.product" },
multicoreSupport = true)
public class SearchContext {

        @Bean
        public SolrServer solrServer(@Value("${solr.host}") String solrHost) {
                return new HttpSolrServer(solrHost);
        }

}
```

File: WebContext.java
```
import java.util.List;
import org.springframework.context.annotation.Bean;
import org.springframework.context.annotation.Configuration;
import org.springframework.data.web.PageableHandlerMethodArgumentResolver;
import org.springframework.web.method.support.HandlerMethodArgumentResolver;
import org.springframework.web.servlet.config.annotation.ViewControllerRegistry;
import org.springframework.web.servlet.config.annotation.WebMvcConfigurerAdapter;


/**
 * @author kkoitzsch
 */
@Configuration
public class WebContext {
        @Bean
```

```
        public WebMvcConfigurerAdapter mvcViewConfigurer() {
                return new WebMvcConfigurerAdapter() {
                        @Override
                        public void addViewControllers(ViewControllerRegistry registry) {

                                registry.addViewController("/").setViewName("search");
                                registry.addViewController("/monitor").setViewName("monitor");
                        }
                        @Override
                        public void addArgumentResolvers(List<HandlerMethodArgumentResolver>
                           argumentResolvers) {
                                argumentResolvers.add(new
                                    PageableHandlerMethodArgumentResolver());
                        }
                };
        }
}
```

Listing 6-5. Spring Data code example using Solr

```
public static void main(String[] args) throws IOException {
        String text = "The World is a great place";
        Properties props = new Properties();
        props.setProperty("annotators", "tokenize, ssplit, pos, lemma, parse, sentiment");
        StanfordCoreNLP pipeline = new StanfordCoreNLP(props);

        Annotation annotation = pipeline.process(text);
        List<CoreMap> sentences = annotation.get(CoreAnnotations.SentencesAnnotation.class);
        for (CoreMap sentence : sentences) {
            String sentiment = sentence.get(SentimentCoreAnnotations.SentimentClass.class);
            System.out.println(sentiment + "\t" + sentence);
        }
    }
```

Listing 6-6. Spring Data code example using Elasticsearch (unit test)

```
package com.apress.probda.search.elasticsearch;
import com.apress.probda.search.elasticsearch .Application;
import com.apress.probda.search.elasticsearch .Post;
import com.apress.probda.search.elasticsearch.Tag;
import com.apress.probda.search.elasticsearch.PostService;
import org.junit.Before;
import org.junit.Test;
import org.junit.runner.RunWith;
import org.springframework.beans.factory.annotation.Autowired;
import org.springframework.boot.test.SpringApplicationConfiguration;
import org.springframework.data.domain.Page;
import org.springframework.data.domain.PageRequest;
import org.springframework.data.elasticsearch.core.ElasticsearchTemplate;
import org.springframework.test.context.junit4.SpringJUnit4ClassRunner;
```

```java
import java.util.Arrays;
import static org.hamcrest.CoreMatchers.notNullValue;
import static org.hamcrest.core.Is.is;
import static org.junit.Assert.assertThat;


@RunWith(SpringJUnit4ClassRunner.class)
@SpringApplicationConfiguration(classes = Application.class)
public class PostServiceImplTest{
    @Autowired
    private PostService postService;
    @Autowired
    private ElasticsearchTemplate elasticsearchTemplate;

    @Before
    public void before() {
        elasticsearchTemplate.deleteIndex(Post.class);
        elasticsearchTemplate.createIndex(Post.class);
        elasticsearchTemplate.putMapping(Post.class);
        elasticsearchTemplate.refresh(Post.class, true);
    }
    //@Test
    public void testSave() throws Exception {
        Tag tag = new Tag();
        tag.setId("1");
        tag.setName("tech");
        Tag tag2 = new Tag();
        tag2.setId("2");
        tag2.setName("elasticsearch");
        Post post = new Post();
        post.setId("1");
        post.setTitle("Bigining with spring boot application and elasticsearch");
        post.setTags(Arrays.asList(tag, tag2));
        postService.save(post);
        assertThat(post.getId(), notNullValue());
        Post post2 = new Post();
        post2.setId("1");
        post2.setTitle("Bigining with spring boot application");
        post2.setTags(Arrays.asList(tag));
        postService.save(post);
        assertThat(post2.getId(), notNullValue());
    }
    public void testFindOne() throws Exception {
    }

    public void testFindAll() throws Exception {
    }

    @Test
    public void testFindByTagsName() throws Exception {
        Tag tag = new Tag();
        tag.setId("1");
```

```
        tag.setName("tech");
        Tag tag2 = new Tag();
        tag2.setId("2");
        tag2.setName("elasticsearch");

        Post post = new Post();
        post.setId("1");
        post.setTitle("Bigining with spring boot application and elasticsearch");
        post.setTags(Arrays.asList(tag, tag2));
        postService.save(post);

        Post post2 = new Post();
        post2.setId("1");
        post2.setTitle("Bigining with spring boot application");
        post2.setTags(Arrays.asList(tag));
        postService.save(post);

        Page<Post> posts  = postService.findByTagsName("tech", new PageRequest(0,10));
        Page<Post> posts2 = postService.findByTagsName("tech", new PageRequest(0,10));
        Page<Post> posts3 = postService.findByTagsName("maz", new PageRequest(0,10));

      assertThat(posts.getTotalElements(), is(1L));
       assertThat(posts2.getTotalElements(), is(1L));
       assertThat(posts3.getTotalElements(), is(0L));
    }
}
```

6.7   Using LingPipe and GATE for Customized Search

In this section, we will review a pair of useful analytical tools which may be used with Lucene and Solr to enhance natural language processing (NLP) analytics capabilities in a distributed analytic application. LingPipe (http://alias-i.com/lingpipe/) and GATE (General Architecture for Text Engineering, https://gate.ac.uk) can be used to add natural language processing capabilities to analytical systems. A typical architecture for an NLP based analytical system might be similar to Figure 6-22.

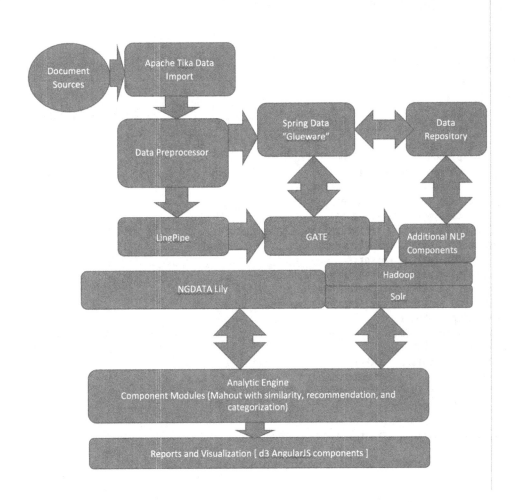

Figure 6-22. *NLP system architecture, using LingPipe, GATE, and NGDATA Lily*

Natural language processing systems can be designed and built in a similar fashion to any other distributed pipelining system. The only difference is the necessary adjustments for the particular nature of the data and metadata itself. LingPipe, GATE, Vowpal Wabbit, and StanfordNLP allow for the processing, parsing, and "understanding" of text, and packages such as Emir/Caliph, ImageTerrier, and HIPI provide features to analyze and index image- and signal-based data. You may also wish to add packages to help with geolocation, such as SpatialHadoop (`http://spatialhadoop.cs.umn.edu`), which is discussed in more detail in Chapter 14.

Various input formats including raw text, XML, HTML, and PDF documents can be processed by GATE, as well as relational data/JDBC-mediated data. This includes data imported from Oracle, PostgreSQL, and others.

The Apache Tika import component might be implemented as in Listing 6-7.

Listing 6-7. Apache Tika import routines for use throughout the PROBDA System

```
Package com.apress.probda.io;

import java.io.*;
import java.nio.file.Paths;
import java.util.ArrayList;
import java.util.List;
import java.util.Map;
import java.util.Set;

import com.apress.probda.pc.AbstractProbdaKafkaProducer;
import org.apache.commons.lang3.StringUtils;
import org.apache.tika.exception.TikaException;
import org.apache.tika.io.TikaInputStream;
import org.apache.tika.metadata.Metadata;
import org.apache.tika.metadata.serialization.JsonMetadata;
import org.apache.tika.parser.ParseContext;
import org.apache.tika.parser.Parser;
import org.apache.tika.parser.isatab.ISArchiveParser;
import org.apache.tika.sax.ToHTMLContentHandler;
import org.dia.kafka.solr.consumer.SolrKafkaConsumer;
import org.json.simple.JSONArray;
import org.json.simple.JSONObject;
import org.json.simple.JSONValue;
import org.json.simple.parser.JSONParser;
import org.json.simple.parser.ParseException;
import org.xml.sax.ContentHandler;
import org.xml.sax.SAXException;

import static org.dia.kafka.Constants.*;

public class ISAToolsKafkaProducer extends AbstractKafkaProducer {

    /**
     * Tag for specifying things coming out of LABKEY
     */
    public final static String ISATOOLS_SOURCE_VAL = "ISATOOLS";
    /**
     * ISA files default prefix
     */
    private static final String DEFAULT_ISA_FILE_PREFIX = "s_";
    /**
     * Json jsonParser to decode TIKA responses
     */
    private static JSONParser jsonParser = new JSONParser();
    ;

    /**
     * Constructor
     */
    public ISAToolsKafkaProducer(String kafkaTopic, String kafkaUrl) {
```

```java
        initializeKafkaProducer(kafkaTopic, kafkaUrl);
    }

    /**
     * @param args
     */
    public static void main(String[] args) throws IOException {
        String isaToolsDir = null;
        long waitTime = DEFAULT_WAIT;
        String kafkaTopic = KAFKA_TOPIC;
        String kafkaUrl = KAFKA_URL;

        // TODO Implement commons-cli
        String usage = "java -jar ./target/isatools-producer.jar [--tikaRESTURL <url>]
[--isaToolsDir <dir>] [--wait <secs>] [--kafka-topic <topic_name>] [--kafka-url]\n";

        for (int i = 0; i < args.length - 1; i++) {
            if (args[i].equals("--isaToolsDir")) {
                isaToolsDir = args[++i];
            } else if (args[i].equals("--kafka-topic")) {
                kafkaTopic = args[++i];
            } else if (args[i].equals("--kafka-url")) {
                kafkaUrl = args[++i];
            }
        }

        // Checking for required parameters
        if (StringUtils.isEmpty(isaToolsDir)) {
            System.err.format("[%s] A folder containing ISA files should be specified.\n",
              ISAToolsKafkaProducer.class.getSimpleName());
            System.err.println(usage);
            System.exit(0);
        }

        // get KafkaProducer
        final ISAToolsKafkaProducer isatProd = new ISAToolsKafkaProducer(kafkaTopic, kafkaUrl);
        DirWatcher dw = new DirWatcher(Paths.get(isaToolsDir));

        // adding shutdown hook for shutdown gracefully
        Runtime.getRuntime().addShutdownHook(new Thread(new Runnable() {
            public void run() {
                System.out.println();
                System.out.format("[%s] Exiting app.\n", isatProd.getClass().getSimpleName());
                isatProd.closeProducer();
            }
        }));

        // get initial ISATools files
        List<JSONObject> newISAUpdates = isatProd.initialFileLoad(isaToolsDir);
```

```java
        // send new studies to kafka
        isatProd.sendISAToolsUpdates(newISAUpdates);
        dw.processEvents(isatProd);

    }

    /**
     * Checks for files inside a folder
     *
     * @param innerFolder
     * @return
     */
    public static List<String> getFolderFiles(File innerFolder) {
        List<String> folderFiles = new ArrayList<String>();
        String[] innerFiles = innerFolder.list(new FilenameFilter() {
            public boolean accept(File dir, String name) {
                if (name.startsWith(DEFAULT_ISA_FILE_PREFIX)) {
                    return true;
                }
                return false;
            }
        });

        for (String innerFile : innerFiles) {
            File tmpDir = new File(innerFolder.getAbsolutePath() + File.separator + innerFile);
            if (!tmpDir.isDirectory()) {
                folderFiles.add(tmpDir.getAbsolutePath());
            }
        }
        return folderFiles;
    }

    /**
     * Performs the parsing request to Tika
     *
     * @param files
     * @return a list of JSON objects.
     */
    public static List<JSONObject> doTikaRequest(List<String> files) {
        List<JSONObject> jsonObjs = new ArrayList<JSONObject>();

        try {
            Parser parser = new ISArchiveParser();
            StringWriter strWriter = new StringWriter();

            for (String file : files) {
                JSONObject jsonObject = new JSONObject();

                // get metadata from tika
                InputStream stream = TikaInputStream.get(new File(file));
```

```
                ContentHandler handler = new ToHTMLContentHandler();
                Metadata metadata = new Metadata();
                ParseContext context = new ParseContext();
                parser.parse(stream, handler, metadata, context);

                // get json object
                jsonObject.put(SOURCE_TAG, ISATOOLS_SOURCE_VAL);
                JsonMetadata.toJson(metadata, strWriter);
                jsonObject = adjustUnifiedSchema((JSONObject) jsonParser.parse(new
                            String(strWriter.toString())));
                //TODO Tika parsed content is not used needed for now
                //jsonObject.put(X_TIKA_CONTENT, handler.toString());
                System.out.format("[%s] Tika message: %s \n", ISAToolsKafkaProducer.class.
                    getSimpleName(), jsonObject.toJSONString());

                jsonObjs.add(jsonObject);

                strWriter.getBuffer().setLength(0);
            }
            strWriter.flush();
            strWriter.close();

    } catch (IOException e) {
        e.printStackTrace();
    } catch (ParseException e) {
        e.printStackTrace();
    } catch (SAXException e) {
        e.printStackTrace();
    } catch (TikaException e) {
        e.printStackTrace();
    }
    return jsonObjs;
}


private static JSONObject adjustUnifiedSchema(JSONObject parse) {
    JSONObject jsonObject = new JSONObject();
    List invNames = new ArrayList<String>();
    List invMid = new ArrayList<String>();
    List invLastNames = new ArrayList<String>();

    Set<Map.Entry> set = parse.entrySet();
    for (Map.Entry entry : set) {
        String jsonKey = SolrKafkaConsumer.updateCommentPreffix(entry.getKey().toString());
        String solrKey = ISA_SOLR.get(jsonKey);

//          System.out.println("solrKey " + solrKey);
        if (solrKey != null) {
//              System.out.println("jsonKey: " + jsonKey + " -> solrKey: " + solrKey);
            if (jsonKey.equals("Study_Person_First_Name")) {
```

```java
                    invNames.addAll(((JSONArray) JSONValue.parse(entry.getValue().
                        toString()))));
                } else if (jsonKey.equals("Study_Person_Mid_Initials")) {
                    invMid.addAll(((JSONArray) JSONValue.parse(entry.getValue().
                        toString()))));
                } else if (jsonKey.equals("Study_Person_Last_Name")) {
                    invLastNames.addAll(((JSONArray) JSONValue.parse(entry.getValue().
                        toString()))));
                }
                jsonKey = solrKey;
            } else {
                jsonKey = jsonKey.replace(" ", "_");
            }
            jsonObject.put(jsonKey, entry.getValue());
        }

        JSONArray jsonArray = new JSONArray();

        for (int cnt = 0; cnt < invLastNames.size(); cnt++) {
            StringBuilder sb = new StringBuilder();
            if (!StringUtils.isEmpty(invNames.get(cnt).toString()))
                sb.append(invNames.get(cnt)).append(" ");
            if (!StringUtils.isEmpty(invMid.get(cnt).toString()))
                sb.append(invMid.get(cnt)).append(" ");
            if (!StringUtils.isEmpty(invLastNames.get(cnt).toString()))
                sb.append(invLastNames.get(cnt));
            jsonArray.add(sb.toString());
        }
        if (!jsonArray.isEmpty()) {
            jsonObject.put("Investigator", jsonArray.toJSONString());
        }
        return jsonObject;
    }


    /**
     * Send message from IsaTools to kafka
     *
     * @param newISAUpdates
     */
    void sendISAToolsUpdates(List<JSONObject> newISAUpdates) {
        for (JSONObject row : newISAUpdates) {
            row.put(SOURCE_TAG, ISATOOLS_SOURCE_VAL);
            this.sendKafka(row.toJSONString());
            System.out.format("[%s] New message posted to kafka.\n", this.getClass().
                getSimpleName());
        }
    }
}
```

```java
/**
 * Gets the application updates from a directory
 *
 * @param isaToolsTopDir
 * @return
 */
private List<JSONObject> initialFileLoad(String isaToolsTopDir) {
    System.out.format("[%s] Checking in %s\n", this.getClass().getSimpleName(),
        isaToolsTopDir);
    List<JSONObject> jsonParsedResults = new ArrayList<JSONObject>();
    List<File> innerFolders = getInnerFolders(isaToolsTopDir);

    for (File innerFolder : innerFolders) {
        jsonParsedResults.addAll(doTikaRequest(getFolderFiles(innerFolder)));
    }

    return jsonParsedResults;
}

/**
 * Gets the inner folders inside a folder
 *
 * @param isaToolsTopDir
 * @return
 */
private List<File> getInnerFolders(String isaToolsTopDir) {
    List<File> innerFolders = new ArrayList<File>();
    File topDir = new File(isaToolsTopDir);
    String[] innerFiles = topDir.list();
    for (String innerFile : innerFiles) {
        File tmpDir = new File(isaToolsTopDir + File.separator + innerFile);
        if (tmpDir.isDirectory()) {
            innerFolders.add(tmpDir);
        }
    }
    return innerFolders;
}
}
```

INSTALLING AND TESTING LINGPIPE, GATE, AND STANFORD CORE NLP

1. First install LingPipe by downloading the LingPipe release JAR file from `http://alias-i.com/lingpipe/web/download.html`. You may also download LingPipe models that interest you from `http://alias-i.com/lingpipe/web/models.html`. Follow the directions so as to place the models in the correct directory so that LingPipe may pick up the models for the appropriate demos which require them.

2. Download GATE from University of Sheffield web site (`https://gate.ac.uk`), and use the installer to install GATE components. The installation dialog is quite easy to use and allows you to selectively install a variety of components, as shown in Figure 6-24.

3. We will also introduce the StanfordNLP (`http://stanfordnlp.github.io/CoreNLP/#human-languages-supported`) library component for our example.

   To get started with Stanford NLP, download the CoreNLP zip file from the GitHub link above. Expand the zip file.

   Make sure the following dependencies are in your pom.xml file:

```
<dependency>
  <groupId>edu.stanford.nlp</groupId>
  <artifactId>stanford-corenlp</artifactId>
  <version>3.5.2</version>
  <classifier>models</classifier>
</dependency>
<dependency>
  <groupId>edu.stanford.nlp</groupId>
  <artifactId>stanford-corenlp</artifactId>
  <version>3.5.2</version>
</dependency>
<dependency>
  <groupId>edu.stanford.nlp</groupId>
  <artifactId>stanford-parser</artifactId>
  <version>3.5.2</version>
</dependency>
```

   Go to Stanford NLP "home directory" (where the pom.xml file is located) and do

```
mvn clean package
```

   then test the interactive NLP shell to insure correct behavior. Type

```
./corenlp.sh
```

   to start the interactive NLP shell. Type some sample text into the shell to see the parser in action. The results shown will be similar to those shown in Figure 6-17.

```
● ● ●                    ■ stanford-corenlp-full-2015-12-09 — java — 80×39                    ■

Entering interactive shell. Type q RETURN or EOF to quit.
NLP> now is the winter of our discontent
Sentence #1 (7 tokens):
now is the winter of our discontent
[Text=now CharacterOffsetBegin=0 CharacterOffsetEnd=3 PartOfSpeech=RB Lemma=now
NamedEntityTag=DATE NormalizedNamedEntityTag=PRESENT_REF Timex=<TIMEX3 tid="t1"
type="DATE" value="PRESENT_REF">now</TIMEX3>]
[Text=is CharacterOffsetBegin=4 CharacterOffsetEnd=6 PartOfSpeech=VBZ Lemma=be N
amedEntityTag=O]
[Text=the CharacterOffsetBegin=7 CharacterOffsetEnd=10 PartOfSpeech=DT Lemma=the
 NamedEntityTag=O]
[Text=winter CharacterOffsetBegin=11 CharacterOffsetEnd=17 PartOfSpeech=NN Lemma
=winter NamedEntityTag=DATE NormalizedNamedEntityTag=XXXX-WI Timex=<TIMEX3 tid="
t2" type="DATE" value="XXXX-WI">winter</TIMEX3>]
[Text=of CharacterOffsetBegin=18 CharacterOffsetEnd=20 PartOfSpeech=IN Lemma=of
NamedEntityTag=O]
[Text=our CharacterOffsetBegin=21 CharacterOffsetEnd=24 PartOfSpeech=PRP$ Lemma=
we NamedEntityTag=O]
[Text=discontent CharacterOffsetBegin=25 CharacterOffsetEnd=35 PartOfSpeech=NN L
emma=discontent NamedEntityTag=O]
(ROOT
  (SINV
    (ADVP (RB now))
    (VP (VBZ is))
    (NP
      (NP (DT the) (NN winter))
      (PP (IN of)
        (NP (PRP$ our) (NN discontent))))))

root(ROOT-0, is-2)
advmod(is-2, now-1)
det(winter-4, the-3)
nsubj(is-2, winter-4)
case(discontent-7, of-5)
nmod:poss(discontent-7, our-6)
nmod:of(winter-4, discontent-7)

NLP> []
```

Figure 6-23. *StanfordNLP interactive shell in action*

We can define an interface for generalized search as follows:

Listing 6-8. ProbdaSearchEngine java interface stub

```
public interface ProbdaSearchEngine<T> {

  <Q> List<T> search(final String field, final Q query, int maximumResultCount);

  List<T> search(final String query, int maximumResultCount);
……………}
```

Two different method signatures for search() are present. One is specifically for the field and query combination. Query is the Lucene query as a string, and maximumResultCount limits the number of result elements to a manageable amount.

We can define the implementation of the ProbdaSearchEngine interface as in Listing 6-8.

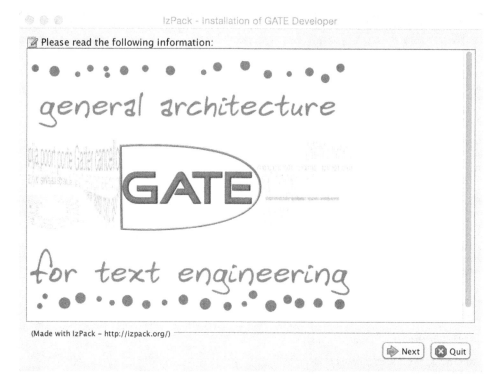

Figure 6-24. GATE Installation dialog. GATE is easy to install and use.

Simply click through the installation wizard. Refer to the web site and install all software components offered.

To use LingPipe and GATE in a program, let's work through a simple example, as shown in Listing 6-9. Please refer to some of the references at the end of the chapter to get a more thorough overview of the features that LingPipe and GATE can provide.

Listing 6-9. LingPipe | GATE | StanfordNLP Java test program, imports

```
package com.apress.probda.nlp;

import java.io.*;
import java.util.*;

import edu.stanford.nlp.io.*;
import edu.stanford.nlp.ling.*;
import edu.stanford.nlp.pipeline.*;
import edu.stanford.nlp.trees.*;
import edu.stanford.nlp.util.*;

public class ProbdaNLPDemo {
```

```java
public static void main(String[] args) throws IOException {
  PrintWriter out;
  if (args.length > 1) {
    out = new PrintWriter(args[1]);
  } else {
    out = new PrintWriter(System.out);
  }
  PrintWriter xmlOut = null;
  if (args.length > 2) {
    xmlOut = new PrintWriter(args[2]);
  }

  StanfordCoreNLP pipeline = new StanfordCoreNLP();
  Annotation annotation;
  if (args.length > 0) {
    annotation = new Annotation(IOUtils.slurpFileNoExceptions(args[0]));
  } else {
    annotation = new Annotation("No reply from local Probda email site");
  }

  pipeline.annotate(annotation);
  pipeline.prettyPrint(annotation, out);
  if (xmlOut != null) {
    pipeline.xmlPrint(annotation, xmlOut);
  }
  List<CoreMap> sentences = annotation.get(CoreAnnotations.SentencesAnnotation.class);
  if (sentences != null && sentences.size() > 0) {
    CoreMap sentence = sentences.get(0);
    Tree tree = sentence.get(TreeCoreAnnotations.TreeAnnotation.class);
    out.println();
    out.println("The first sentence parsed is:");
    tree.pennPrint(out);
  }
}

}
```

6.8   Summary

In this chapter, we took a quick overview of the Apache Lucene and Solr ecosystem. Interestingly, although Hadoop and Solr started out together as part of the Lucene ecosystem, they have since gone their separate ways and evolved into useful independent frameworks. This doesn't mean that the Solr and Hadoop ecosystems cannot work together very effectively, however. Many Apache components, such as Mahout, LingPipe, GATE, and Stanford NLP, work seamlessly with Lucene and Solr. New technology additions to Solr, such as SolrCloud and others, make it easier to use RESTful APIs to interface to the Lucene/Solr technologies.

We worked through a complete example of using Solr and its ecosystem: from downloading, massaging, and inputting the data set to transforming the data and outputting results in a variety of data formats. It becomes even more clear that Apache Tika and Spring Data are extremely useful for data pipeline "glue."

We did not neglect competitors to the Lucene/Solr technology stack. We were able to discuss Elasticsearch, a strong alternative to Lucene/Solr, and describe some of the pros and cons of using Elasticsearch over a more "vanilla Lucene/Solr" approach. One of the most interesting parts of Elasticsearch is the seamless ability to visualize data, as we showed while exploring the crime statistics of Sacramento.

In the next chapter, we will discuss a number of analytic techniques and algorithms which are particularly useful for building distributed analytical systems, building upon what we've learned so far.

6.9   References

Awad, Mariette and Khanna, Rahul. *Efficient Learning Machines*. New York: Apress Open Publications, 2015.

Babenko, Dmitry and Marmanis,Haralambos. *Algorithms of the Intelligent Web*. Shelter Island : Manning Publications, 2009.

Guller, Mohammed. *Big Data Analytics with Apache Spark*. New York: Apress Press, 2015.

Karambelkar, Hrishikesh. *Scaling Big Data with Hadoop and Solr*. Birmingham, UK: PACKT Publishing, 2013.

Konchady, Manu. *Building Search Applications: Lucene, LingPipe and GATE*. Oakton, VA : Mustru Publishing, 2008.

Mattmann, Chris A. and Zitting, Jukka I. *Tika in Action*. Shelter Island: Manning Publications, 2012.

Pollack, Mark, Gierke, Oliver, Risberg, Thomas, Brisbin, Jon, Hunger, Michael. *Spring Data: Modern Data Access for Enterprise Java*. Sebastopol, CA: O'Reilly Media, 2012.

Venner, Jason. *Pro Hadoop*. New York NY: Apress Press, 2009.

PART II

Architectures and Algorithms

The second part of our book discusses standard architectures, algorithms, and techniques to build analytic systems using Hadoop. We also investigate rule-based systems for control, scheduling, and system orchestration and showcase how a rule-based controller can be a natural adjunct to a Hadoop-based analytical system.

■ ■ ■

CHAPTER 7

An Overview of Analytical Techniques and Algorithms

In this chapter, we provide an overview of four categories of algorithm: statistical, Bayesian, ontology-driven, and hybrid algorithms which leverage the more basic algorithms found in standard libraries to perform more in-depth and accurate analyses using Hadoop.

7.1 Survey of Algorithm Types

It turns out that Apache Mahout and most of the other mainstream machine learning toolkits support a wide range of the algorithms we're interested in. For example, see Figure 7-1 for a survey of the algorithms supported by Apache Mahout.

Number	Algorithm Name	Algorithm Type	Description
1	naïve Bayes	classifier	simple Bayesian classifier: present in almost all modern toolkits
2	hidden Markov model	classifier	system state prediction by outcome observation
3	(learning) random forest	classifier	Random forest algorithms (sometimes known as random decision forests) are an ensemble learning method for classification, regression, and other tasks, that construct a collection of decision trees at training time, outputting the class that is the mode of the classification classes or mean prediction (regression) of the individual trees.
4	(learning) multilayer perceptron (LMP)	classifier	also implemented in the Theano toolkit and several others.
5	(learning) logistic regression	classifier	also supported in scikit-learn. Really a technique for classification, not regression.
6	stochastic gradient descent (SGD)	optimizer, model finding	an objective function minimization routine also supported in H2O and Vowpal Wabbit, among others

(continued)

© Kerry Koitzsch 2017

K. Koitzsch, *Pro Hadoop Data Analytics*, DOI 10.1007/978-1-4842-1910-2_7

Number	Algorithm Name	Algorithm Type	Description
7	genetic algorithms (GA)	genetic algorithm	According to Wikipedia, "In the field of mathematical optimization, a genetic algorithm (GA) is a search heuristic that mimics the process of natural selection. This heuristic (also sometimes called a meta-heuristic) is routinely used to generate useful solutions to optimization and search problems."
8	singular value decomposition (SVD)	dimensionality reduction	matrix decomposition for dimensionality reduction
9	collaborative filtering (CF)	recommender	a technique used by some recommender systems
10	latent Dirichlet allocation (LDA)	topic modeler	a powerful algorithm (learner) which automatically (and jointly) clusters words into "topics" as well as clustering documents into topic "mixtures"
11	spectral clustering	clusterer	
12	frequent pattern mining	data miner	
13	k-means Clustering	clusterer	ordinary and fuzzy k-means are available using Mahout
14	canopy clustering	clusterer	preprocessing step for k-means clusterer: two-threshold system

Statistical and numerical algorithms are the most straightforward type of distributed algorithm we can use. Statistical techniques include the use of standard statistic computations such as those shown in Figure 7-1.

$$\mu = \frac{1}{n} \sum_{i=1}^{n} x_i \qquad \text{Mean}$$

$$\sigma = \left[\frac{1}{n-1} \sum_{i=1}^{n} \left(x_i - \mu \right)^2 \right]^{0.5} \qquad \text{Standard deviation}$$

$$f(x) = \frac{1}{\sqrt{2\pi}\sigma} e^{-\frac{(x-\mu)^2}{2\sigma^2}} \qquad \text{Normal distribution}$$

Figure 7-1. *Mean, standard deviation, and normal distribution are often used in statistical methods*

Bayesian techniques are one of the most effective techniques for building classifiers, data modeling, and other purposes.

Ontology-driven algorithms, on the other hand, are a whole family of algorithms that rely on logical, structured, hierarchical modeling, grammars, and other techniques to provide infrastructure for modeling, data mining, and drawing inferences about data sets.

Hybrid algorithms combine one or more modules consisting of different types of algorithm, linked together with glueware, to provide a more flexible and powerful data pipeline than would be possible with only a single algorithm type. For example, a neural net technology may be combined with a Bayesian technology and an ML technology to create "learning Bayesian networks," a very interesting example of the synergy that can be obtained by using a hybrid approach.

7.2   Statistical / Numerical Techniques

Statistical classes and support methods in the example system are found in the `com.apress.probda.algorithms.statistics` subpackage.

We can see a simple distributed technology stack using Apache Storm in Figure 7-2.

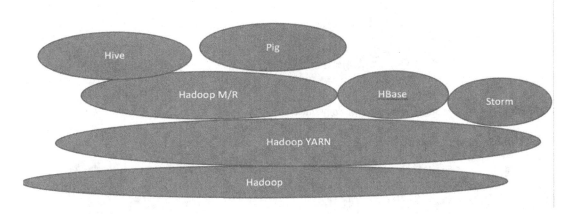

Figure 7-2. *A distributed technology stack including Apache Storm*

We can see a Tachyon-centric technology stack in Figure 7-4. Tachyon is a fault tolerant distributed in-memory file system

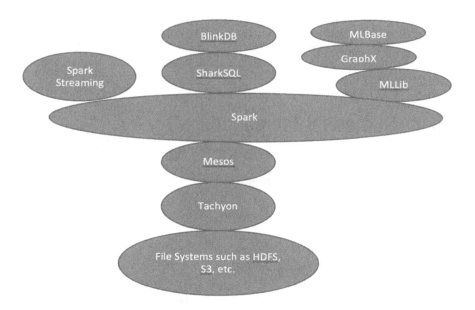

Figure 7-3. *An Apache Spark-centric technology stack*

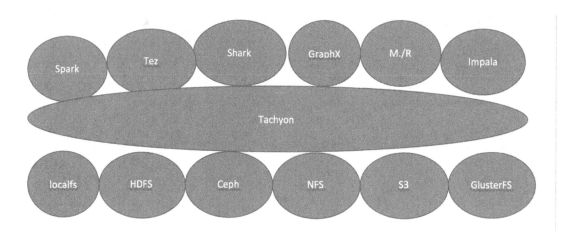

Figure 7-4. *A Tachyon-centric technology stack, showing some of the associated ecosystem*

7.3 Bayesian Techniques

The Bayesian techniques we implement in the example system are found in the package com.prodba. algorithms.bayes.

Some of the Bayesian techniques (besides the naïve Bayes algorithm) supported by our most popular libraries include the ones shown in Figure 7-1.

The naïve Bayesian classifier is based upon the Fundamental Bayes equation as shown in Figure 7-5.

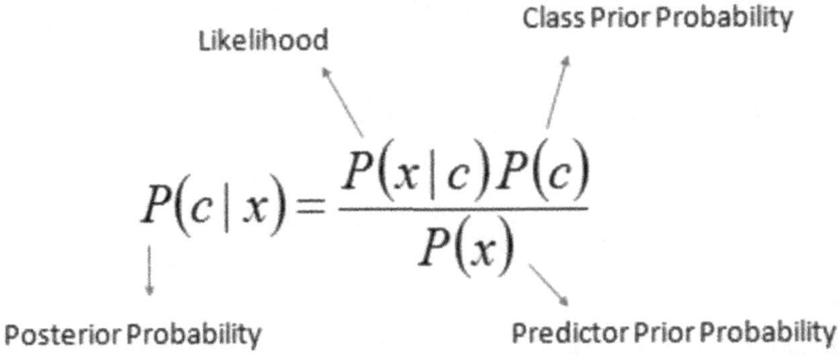

$$P(c \mid x) = \frac{P(x \mid c)P(c)}{P(x)}$$

Figure 7-5. *The fundamental Bayes equation*

The equation contains four main probability types: posterior probability, likelihood, class prior probability, and predictor prior probability. These terms are explained in the references at the end of the chapter.

We can try out the Mahout text classifier in a straightforward way. First, download one of the basic data sets to test with.

7.4   Ontology Driven Algorithms

The ontology driven components and support classes are to be found in the `com.apress.probda.algorithms.ontology` subpackage.

To include the Protégé Core component, add the following Maven dependency to your project pom.xml file.

```
<dependency>
        <groupId>edu.stanford.protege</groupId>
        <artifactId>protege-common</artifactId>
        <version>5.0.0-beta-24</version>
</dependency>
```

Register and download Protégé from the web site:

`http://protege.stanford.edu/products.php#desktop-protégé.`

Ontologies may be defined interactively by using an ontology editor such as Stanford's Protégé system, as shown in Figure 7-5.

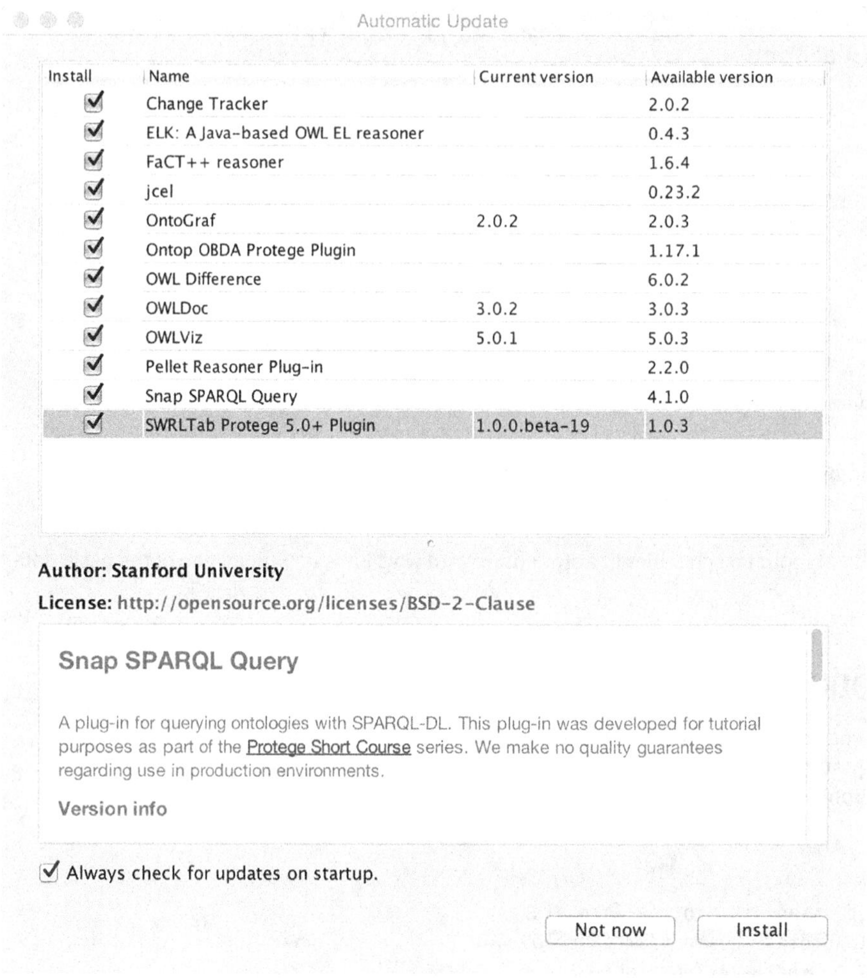

Automatic Update

Install	Name	Current version	Available version
☑	Change Tracker		2.0.2
☑	ELK: A Java-based OWL EL reasoner		0.4.3
☑	FaCT++ reasoner		1.6.4
☑	jcel		0.23.2
☑	OntoGraf	2.0.2	2.0.3
☑	Ontop OBDA Protege Plugin		1.17.1
☑	OWL Difference		6.0.2
☑	OWLDoc	3.0.2	3.0.3
☑	OWLViz	5.0.1	5.0.3
☑	Pellet Reasoner Plug-in		2.2.0
☑	Snap SPARQL Query		4.1.0
☑	SWRLTab Protege 5.0+ Plugin	1.0.0.beta-19	1.0.3

Author: Stanford University
License: http://opensource.org/licenses/BSD-2-Clause

Snap SPARQL Query

A plug-in for querying ontologies with SPARQL-DL. This plug-in was developed for tutorial purposes as part of the **Protege Short Course** series. We make no quality guarantees regarding use in production environments.

Version info

☑ Always check for updates on startup.

[Not now]   [Install]

Figure 7-6. Setting up SPARQL functionality with the Stanford toolkit interactive setup

You can safely select all the components, or just the ones you need. Refer to the individual online documentation pages to see if the components are right for your application.

Figure 7-7. *Using an ontology editor to define ontologies, taxonomies, and grammars*

7.5   Hybrid Algorithms: Combining Algorithm Types

The hybrid algorithms implemented in the Probda example system are found in the com.apress.prodbda. algorithms.hybrid subpackage.

We can mix and match algorithm types to construct better data pipelines. These "hybrid systems" might be somewhat more complex—of necessity they usually have several additional components—but they make up for it in usefulness.

One of the most effective kinds of hybrid algorithm is that of the so-called "deep learning" component. Not everyone considers deep learners as a hybrid algorithm (they are essentially, in most cases, built on multilayer neural net technology), but there are some compelling reasons to treat deep learners as hybrid systems, as we will discuss below.

So-called "deep learning" techniques include those shown in Figure yy-yy. DeepLearning4J and TensorFlow toolkit are two of the more popular and powerful deep learning libraries currently available. Check out TensorFlow at https://www.tensorflow.org/versions/r0.10/get_started/basic_usage. htmlautoenc. Theano is a Python-based multidimensional array library. Check out http://deeplearning. net/tutorial/dA.html for more details about how to use Theano.

Number	Algorithm Name	Algorithm Type	Toolkit	Description
1	Deep Belief Networks	neural net	Deeplearning4j, TensorFlow, Theano	multiple layers of hidden units, with layer interconnectivity only
2	(Stacked, Denoising) Autoencoders (DA)	variations on basic autoencoder principles	Deeplearning4j, TensorFlow, Theano	A stacked autoencoder is a neural net consisting of multiple layers of sparse autoencoders in which each layer's outputs are wired to the successive layers' inputs. Denoising autoencoders can accept a partially corrupted input while recovering the original uncorrupted input.
3	Convolutional Neural Networks (CNN)	neural net, variation of MLP	Deeplearning4j, TensorFlow, Theano	Sparse connectivity and shared weights are two features of CNNs.
4	Long Short-Term Memory Units (LSTM)	recurrent neural net, classifier, predictor	Deeplearning4j, TensorFlow	classification and time series prediction, even sentiment analysis
5	Recurrent Neural Networks	neural net	Deeplearning4j, TensorFlow	classification, time series prediction
6	computation graph	complex network architecture builder	Deeplearning4j, TensorFlow	Computations are represented as graphs.

7.6   Code Examples

In this section we discuss some extended examples of the algorithm types we talked about in earlier sections.

To get a sense of some algorithm comparisons, let's use the movie dataset to evaluate some of the algorithms and toolkits we've talked about.

```
package com.apress.probda.datareader.csv;

import java.io.BufferedReader;
import java.io.BufferedWriter;
import java.io.File;
import java.io.FileNotFoundException;
import java.io.FileOutputStream;
import java.io.FileReader;
import java.io.IOException;
import java.io.OutputStreamWriter;
```

```java
public class FileTransducer {

        /**
         * This routine splits a line which is delimited into fields by the vertical
         * bar symbol '|'
         *
         * @param l
         * @return
         */
        public static String makeComponentsList(String l) {
                String[] genres = l.split("\\|");
                StringBuffer sb = new StringBuffer();
                for (String g : genres) {
                        sb.append("\"" + g + "\",");
                }
                String answer = sb.toString();
                return answer.substring(0, answer.length() - 1);
        }


        /**
         * The main routine processes the standard movie data files so that mahout
         * can use them.
         *
         * @param args
         */
        public static void main(String[] args) {
if (args.length < 4){
System.out.println("Usage: <movie data input><movie output file><ratings input file>
<ratings output file>");
                        System.exit(-1);
                }
                File file = new File(args[0]);
                if (!file.exists()) {
                        System.out.println("File: " + file + " did not exist, exiting...");
                        System.exit(-1);
                }
                System.out.println("Processing file: " + file);
                BufferedWriter bw = null;
                FileOutputStream fos = null;
                String line;
                try (BufferedReader br = new BufferedReader(new FileReader(file))) {
                        int i = 1;
                        File fout = new File(args[1]);
                        fos = new FileOutputStream(fout);
                        bw = new BufferedWriter(new OutputStreamWriter(fos));
                        while ((line = br.readLine()) != null) {
                                String[] components = line.split("::");
                                String number = components[0].trim();
                                String[] titleDate = components[1].split("\\(");
                                String title = titleDate[0].trim();
                                String date = titleDate[1].replace(")", "").trim();
```

```
                                String genreList = makeComponentsList(components[2]);
                                String outLine = "{ \"create\" : { \"_index\" :
\"bigmovie\", \"_type\" : \"film\", \"_id\" : \"" + i
                                        + "\" } }\n" + "{ \"id\": \"" + i + "\",
\"title\" : \"" + title + "\", \"year\":\"" + date
                                        + "\" , \"genre\":[" + genreList + "] }";
                                i++;
                                bw.write(outLine);
                                bw.newLine();
                        }
                } catch (IOException e) {
                        // TODO Auto-generated catch block
                        e.printStackTrace();
                } finally {
                        if (bw != null) {
                                try {
                                        bw.close();
                                } catch (IOException e) {
                                        // TODO Auto-generated catch block
                                        e.printStackTrace();
                                }
                        }
                }
                file = new File(args[2]);
                try (BufferedReader br2 = new BufferedReader(new FileReader(file))) {
                        File fileout = new File(args[3]);
                        fos = new FileOutputStream(fileout);
                        bw = new BufferedWriter(new OutputStreamWriter(fos));
                        while ((line = br2.readLine()) != null) {
                                String lineout = line.replace("::", "\t");
                                bw.write(lineout);
                        }
                } catch (IOException e) {
                        // TODO Auto-generated catch block
                        e.printStackTrace();
                } finally {
                        if (bw != null) {
                                try {
                                        bw.close();
                                } catch (IOException e) {
                                        // TODO Auto-generated catch block
                                        e.printStackTrace();
                                }
                        }
                }
        }
}
```

Execute the following curl command on the command line to import the data set into elastic search:

```
curl -s -XPOST localhost:9200/_bulk --data-binary @index.json; echo
```

Last login: Wed May 4 18:25:07 on ttys006
Kerrys-MBP:~ kerryk$ cd
Kerrys-MBP:~ kerryk$ curl -s -XPOST localhost:9200/_bulk --data-binary @index2.json; echo
{"took":338,"errors":true,"items":[{"create":{"_index":"bigmovie","_type":"film","_id":"1","_version":1,"_shards"
:{"total":2,"successful":1,"failed":0},"status":201}},{"create":{"_index":"bigmovie","_type":"film","_id":"2","_v
ersion":1,"_shards":{"total":2,"successful":1,"failed":0},"status":201}},{"create":{"_index":"bigmovie","_type":"
film","_id":"3","_version":1,"_shards":{"total":2,"successful":1,"failed":0},"status":201}},{"create":{"_index":"
bigmovie","_type":"film","_id":"4","_version":1,"_shards":{"total":2,"successful":1,"failed":0},"status":201}},{"
create":{"_index":"bigmovie","_type":"film","_id":"5","_version":1,"_shards":{"total":2,"successful":1,"failed":0
},"status":201}},{"create":{"_index":"bigmovie","_type":"film","_id":"6","_version":1,"_shards":{"total":2,"succe
ssful":1,"failed":0},"status":201}},{"create":{"_index":"bigmovie","_type":"film","_id":"7","_version":1,"_shards
":{"total":2,"successful":1,"failed":0},"status":201}},{"create":{"_index":"bigmovie","_type":"film","_id":"8","_
version":1,"_shards":{"total":2,"successful":1,"failed":0},"status":201}},{"create":{"_index":
"bigmovie","_type":"film","_id":"10","_version":1,"_shards":{"total":2,"successful":1,"failed":0},"status":201}},
{"create":{"_index":"bigmovie","_type":"film","_id":"11","_version":1,"_shards":{"total":2,"successful":1,"failed
":0},"status":201}},{"create":{"_index":"bigmovie","_type":"film","_id":"12","_version":1,"_shards":{"total":2,"s
uccessful":1,"failed":0},"status":201}},{"create":{"_index":"bigmovie","_type":"film","_id":"13","_version":1,"_s
hards":{"total":2,"successful":1,"failed":0},"status":201}},{"create":{"_index":"bigmovie","_type":"film","_id":"
14","_version":1,"_shards":{"total":2,"successful":1,"failed":0},"status":201}},{"create":{"_index":"bigmovie","_
type":"film","_id":"15","_version":1,"_shards":{"total":2,"successful":1,"failed":0},"status":201}},{"create":{"_
index":"bigmovie","_type":"film","_id":"16","_version":1,"_shards":{"total":2,"successful":1,"failed":0},"status"
:201}},{"create":{"_index":"bigmovie","_type":"film","_id":"17","_version":1,"_shards":{"total":2,"successful":1,
"failed":0},"status":201}},{"create":{"_index":"bigmovie","_type":"film","_id":"18","_version":1,"_shards":{"tota
l":2,"successful":1,"failed":0},"status":201}},{"create":{"_index":"bigmovie","_type":"film","_id":"19","_version
":1,"_shards":{"total":2,"successful":1,"failed":0},"status":201}},{"create":{"_index":"bigmovie","_type":"film",
"_id":"20","_version":1,"_shards":{"total":2,"successful":1,"failed":0},"status":201}},{"create":{"_index":"bigmo
vie","_type":"film","_id":"21","_version":1,"_shards":{"total":2,"successful":1,"failed":0},"status":201}},{"crea
te":{"_index":"bigmovie","_type":"film","_id":"22","_version":1,"_shards":{"total":2,"successful":1,"failed":0},"
status":201}},{"create":{"_index":"bigmovie","_type":"film","_id":"23","_version":1,"_shards":{"total":2,"success
ful":1,"failed":0},"status":201}},{"create":{"_index":"bigmovie","_type":"film","_id":"24","_version":1,"_shards"
:{"total":2,"successful":1,"failed":0},"status":201}},{"create":{"_index":"bigmovie","_type":"film","_id":"25","_
version":1,"_shards":{"total":2,"successful":1,"failed":0},"status":201}},{"create":{"_index":"bigmovie","_type":
"film","_id":"26","_version":1,"_shards":{"total":2,"successful":1,"failed":0},"status":201}},{"create":{"_index"
:"bigmovie","_type":"film","_id":"27","_version":1,"_shards":{"total":2,"successful":1,"failed":0},"status":201}}
,{"create":{"_index":"bigmovie","_type":"film","_id":"28","_version":1,"_shards":{"total":2,"successful":1,"faile
d":0},"status":201}},{"create":{"_index":"bigmovie","_type":"film","_id":"29","_version":1,"_shards":{"total":2,"
successful":1,"failed":0},"status":201}},{"create":{"_index":"bigmovie","_type":"film","_id":"30","_version":1,"_
shards":{"total":2,"successful":1,"failed":0},"status":201}},{"create":{"_index":"bigmovie","_type":"film","_id":
"31","_version":1,"_shards":{"total":2,"successful":1,"failed":0},"status":201}},{"create":{"_index":"bigmovie","
_type":"film","_id":"33","_version":1,"_shards":{"total":2,"successful":1,"failed":0},"status
":201}},{"create":{"_index":"bigmovie","_type":"film","_id":"34","_version":1,"_shards":{"total":2,"successful":1
,"failed":0},"status":201}},{"create":{"_index":"bigmovie","_type":"film","_id":"35","_version":1,"_shards":{"tot
al":2,"successful":1,"failed":0},"status":201}},{"create":{"_index":"bigmovie","_type":"film","_id":"36","_versio
n":1,"_shards":{"total":2,"successful":1,"failed":0},"status":201}},{"create":{"_index":"bigmovie","_type":"film"
,"_id":"37","_version":1,"_shards":{"total":2,"successful":1,"failed":0},"status":201}},{"create":{"_index":"bigm
ovie","_type":"film","_id":"38","_version":1,"_shards":{"total":2,"successful":1,"failed":0},"status":201}},{"cre
ate":{"_index":"bigmovie","_type":"film","_id":"39","_version":1,"_shards":{"total":2,"successful":1,"failed":0},
"status":201}},{"create":{"_index":"bigmovie","_type":"film","_id":"40","_version":1,"_shards":{"total":2,"succes
sful":1,"failed":0},"status":201}},{"create":{"_index":"bigmovie","_type":"film","_id":"41","_version":1,"_shards
":{"total":2,"successful":1,"failed":0},"status":201}},{"create":{"_index":"bigmovie","_type":"film","_id":"42","
_version":1,"_shards":{"total":2,"successful":1,"failed":0},"status":201}},{"create":{"_index
":"bigmovie","_type":"film","_id":"44","_version":1,"_shards":{"total":2,"successful":1,"failed":0},"status":201}
},{"create":{"_index":"bigmovie","_type":"film","_id":"45","_version":1,"_shards":{"total":2,"successful":1,"fail

Figure 7-8. *Importing a standard movie data set example using a CURL command*

Data sets can be imported into Elasticsearch via the command line using a CURL command. Figure 7-8 is the result of executing such a command. The Elasticsearch server returns a JSON data structure which is displayed on the console as well as being indexed into the Elasticsearch system.

Figure 7-9. *Using Kibana as a reporting and visualization tool*

We can see a simple example of using Kibana as a reporting tool in Figure 7-7. Incidentally, we will encounter Kibana and the ELK Stack (Elasticsearch – Logstash – Kibana) throughout much of the remaining content in this book. While there are alternatives to using the ELK stack, it is one of the more painless ways to construct a data analytics system from third-party building blocks.

7.7 Summary

In this chapter, we discussed analytical techniques and algorithms and some criteria for evaluating algorithm effectiveness. We touched on some of the older algorithm types: the statistical and numerical analytical functions. The combination or hybrid algorithm has become particularly important in recent days as techniques from machine learning, statistics, and other areas may be used very effectively in a cooperative way, as we have seen throughout this chapter. For a general introduction to distributed algorithms, see Barbosa (1996).

Many of these algorithm types are extremely complex. Some of them, for example the Bayesian techniques, have a whole literature devoted to them. For a thorough explanation of Bayesian techniques in particular and probabilistic techniques in general, see Zadeh (1992),

In the next chapter, we will discuss rule-based systems, available rule engine systems such as JBoss Drools, and some of the applications of rule-based systems for smart data collection, rule-based analytics, and data pipeline control scheduling and orchestration.

7.8 References

Barbosa, Valmir C. *An Introduction to Distributed Algorithms*. Cambridge, MA: MIT Press, 1996.

Bolstad, William M. *Introduction to Bayesian Statistics*. New York: Wiley Inter-Science, Wiley and Sons, 2004.

Giacomelli, Pico. *Apache Mahout Cookbook*. Birmingham, UK: PACKT Publishing, 2013.

Gupta, Ashish. *Learning Apache Mahout Classification*. Birmingham, UK: PACKT Publishing, 2015.

Marmanis, Haralambos and Babenko, Dmitry. *Algorithms of the Intelligent Web*. Greenwich, CT: Manning Publications, 2009.

Nakhaeizadeh, G. and Taylor, C.C. (eds). *Machine Learning and Statistics: The Interface*. New York: John Wiley and Sons, Inc., 1997.

Parsons, Simon. *Qualitative Methods for Reasoning Under Uncertainty*. Cambridge. MA: MIT Press, 2001.

Pearl, Judea. *Probabilistic Reasoning in Intelligent Systems: Networks of Plausible Inference*. San Mateo, CA: Morgan-Kaufmann Publishers, Inc., 1988.

Zadeh, Lofti A., and Kacprzyk, (eds). *Fuzzy Logic for the Management of Uncertainty*. New York: John Wiley & Sons, Inc., 1992.

■ ■ ■

CHAPTER 8

Rule Engines, System Control, and System Orchestration

In this chapter, we describe the JBoss Drools rule engine and how it may be used to control and orchestrate Hadoop analysis pipelines. We describe an example rule-based controller which can be used for a variety of data types and applications in combination with the Hadoop ecosystem.

■ **Note** Most of the configuration for using the JBoss Drools system is done using Maven dependencies. The appropriate dependencies were shown in Chapter 3 when we discussed the initial setup of JBoss Drools. All the dependencies you need to effectively use JBoss Drools are included in the example PROBDA system available at the code download site.

8.1   Introduction to Rule Systems: JBoss Drools

JBoss Drools (`www.drools.org`) is used throughout the examples in this chapter. It's not the only choice for a rule engine. There are many rule engine frameworks which are freely available, but Drools is a high-powered system that can be used immediately to define many different kinds of control and architecture systems. JBoss Drools has another advantage. There is extensive online and in-print documentation on the Drools system (docs.jboss.org), programming recipes, and details of optimization, as well as explanations of the rule-based technology. Some of the Drools reference books are listed at the end of this chapter. These provide a thorough introduction to rule-based control systems, rule mechanics and editing, and other important details.

In this chapter, we will give a brief overview of rule-based technology with a specific application: defining a complex event processor (CEP) example.

CEPs are a very useful variation on the data pipeline theme, and can be used in practical systems involving everything from credit card fraud detection systems to complex factory control systems.

There are two kinds of data structures at work in all rule systems: rules, of course, which provide the "if-then-else" conditional functionality in a rule-based system (however, we will soon learn that this type of rule, called a "forward chaining" rule, is not the only variety of rule we will encounter; there are also "backward chaining" rules which will be described shortly). An additional data structure used is facts, which are the individual "data items." These are kept in a repository called the working memory store. Please see Figure 8-1 for a simplified view of how this works in the Drools system.

Figure 8-1. *Download the Drools software from the Drools web page*

▪ **Note** This book uses the latest released version of JBoss Drools, which was version 6.4.0 at the time of writing this book. Update the drools.system.version property in your PROBDA project pom.xml if a new version of JBoss Drools is available and you want to use it.

Let's get started by installing JBoss Drools and testing some basic functionality. The installation process is straightforward. From the JBoss Drools homepage, download the current version of Drools by clicking the download button, as shown in Figure 8-1.

cd to the installation directory and run examples/run-examples.sh. You will see a selection menu similar to that in Figure 8-2. Run some output examples to test the Drools system and observe the output in the console, similar to that in Figure 8-3, or a GUI-oriented example, as in Figure 8-4.

Figure 8-2. Select some Drools examples and observe the results to test the Drools system

The built-in Drools examples has a menu from which you can select different test examples, as shown in Figure 8-2. This is a good way to test your overall system set-up and get a sense of what the JBoss Drools system capabilities are.

Figure 8-3. *JBoss Drools GUI-oriented example*

Some of the example components for JBoss Drools have an associated UI, as shown in Figure 8-3.

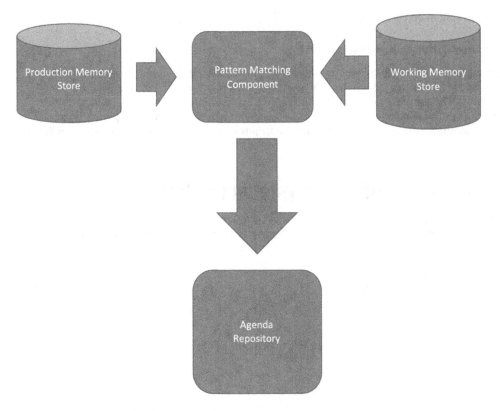

Figure 8-4. *JBoss Drools rule system architecture*

The basic JBoss Drools rule system architecture is shown in Figure 8-4.

■ **Note**　All of the example code found in this system is found in the accompanying example system code base in the Java package com.apress.probda.rulesystem. Please see the associated README file and documentation for additional notes on installation, versioning, and use.

The interface for timestamped Probda events in our system couldn't be easier:

```
package com.probda.rulesystem.cep.model;

import java.util.Date;

public interface IEvent extends Fact {

        public abstract Date getTimestamp();
}
```

The implementation of IEvent looks like this:

Listing 8-1. A basic JBoss Drools program

Let's add a rule system to the evaluation system by way of an example. Simply add the appropriate dependencies for the Drools rule system (Google "drools maven dependencies" for the most up-to-date versions of Drools). The complete pom.xml file (building upon our original) is shown in Listing 3-2. We will be leveraging the functionality of JBoss Drools in a complete analytical engine example in Chapter 8. Please note that we supply dependencies to connect the Drools system with Apache Camel as well as Spring Framework for Drools.

8.2   Rule-based Software Systems Control

Rule-based software systems control can be built up from a scheduling component such as Oozie combined with the appropriate functionalities in JBoss Drools or other rule frameworks, as shown in an example architecture in Figure 8-5.

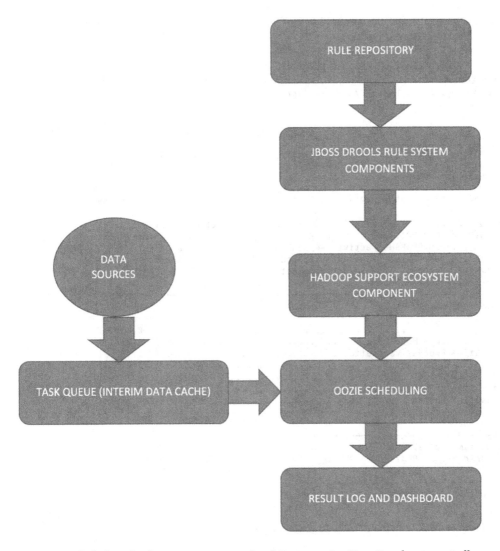

Figure 8-5. *Rule-based software systems control architecture, using JBoss Drools as a controller*

8.3 System Orchestration with JBoss Drools

In this section, we'll discuss a brief example of how to do system orchestration tasks using JBoss Drools as a controller. We will use the Activiti open source project (http://activiti.org) with some examples on how to integrate a workflow orchestrator/controller into a Spring Framework-based project.

```
git clone https://github.com/Activiti/Activiti.git
export ACTIVITI_HOME=/Users/kkoitzsch/activiti
          cd $ACTIVITI_HOME
mvn clean install
```

Don't forget to generate the documentation by
```
        cd $ACTIVITI_HOME/userguide
        mvn install
```
Insure Tomcat is installed. On the Mac platform, do
```
brew install tomcat
```
Tomcat will then be installed at /usr/local/Cellar/tomcat/8.5.3

```
●  ○  ○                    bin — ssh localhost — 80×30
[INFO] --- maven-jar-plugin:2.3.2:jar (default-jar) @ activiti-engine ---
[INFO] Building jar: /Users/kkoitzsch/Activiti/modules/activiti-engine/target/ac
tiviti-engine-5.22.0-SNAPSHOT.jar
[INFO]
[INFO] --- maven-install-plugin:2.4:install (default-install) @ activiti-engine
---
[INFO] Installing /Users/kkoitzsch/Activiti/modules/activiti-engine/target/activ
iti-engine-5.22.0-SNAPSHOT.jar to /Users/kkoitzsch/.m2/repository/org/activiti/a
ctiviti-engine/5.22.0-SNAPSHOT/activiti-engine-5.22.0-SNAPSHOT.jar
[INFO] Installing /Users/kkoitzsch/Activiti/modules/activiti-engine/pom.xml to /
Users/kkoitzsch/.m2/repository/org/activiti/activiti-engine/5.22.0-SNAPSHOT/acti
viti-engine-5.22.0-SNAPSHOT.pom
[INFO] ------------------------------------------------------------------------
[INFO] Reactor Summary:
[INFO]
[INFO] Activiti ........................................... SUCCESS [  0.234 s]
[INFO] Activiti - BPMN Model .............................. SUCCESS [  3.613 s]
[INFO] Activiti - Process Validation ...................... SUCCESS [  1.152 s]
[INFO] Activiti - BPMN Layout ............................. SUCCESS [  1.991 s]
[INFO] Activiti - Image Generator ......................... SUCCESS [  0.787 s]
[INFO] Activiti - BPMN Converter .......................... SUCCESS [  5.121 s]
[INFO] Activiti - Engine .................................. SUCCESS [04:10 min]
[INFO] ------------------------------------------------------------------------
[INFO] BUILD SUCCESS
[INFO] ------------------------------------------------------------------------
[INFO] Total time: 04:27 min
[INFO] Finished at: 2016-07-26T06:36:00-07:00
[INFO] Final Memory: 72M/769M
[INFO] ------------------------------------------------------------------------
Kerrys-MBP:activiti kkoitzsch$ ▓
```

Figure 8-6. *Maven reactor summary for Activiti system install*

Figure 8-6 shows what you can expect from the Maven reactor summary at the end of the Activiti build.

```
export TOMCAT_HOME=/usr/local/Cellar/tomcat/8.5.3
cd $ACTIVITI_HOME/scripts
```

Then run the Activiti script

```
./start-rest-no-jrebel.sh
```

You will see successful startup of Activiti as shown in Figure 8-7.

```
●  ●  ●                    scripts — java ‹ start-rest-no-jrebel.sh — 112×39
07:29:40,145 [localhost-startStop-1] INFO  org.springframework.web.servlet.mvc.method.annotation.RequestMappingH
andlerMapping  - Mapped "{[/runtime/tasks/{taskId}/variables],methods=[POST],params=[],headers=[],consumes=[],pr
oduces=[application/json],custom=[]}" onto public java.lang.Object org.activiti.rest.service.api.runtime.task.Ta
skVariableCollectionResource.createTaskVariable(java.lang.String,javax.servlet.http.HttpServletRequest,javax.ser
vlet.http.HttpServletResponse)
07:29:40,145 [localhost-startStop-1] INFO  org.springframework.web.servlet.mvc.method.annotation.RequestMappingH
andlerMapping  - Mapped "{[/runtime/tasks/{taskId}/variables],methods=[DELETE],params=[],headers=[],consumes=[],
produces=[],custom=[]}" onto public void org.activiti.rest.service.api.runtime.task.TaskVariableCollectionResour
ce.deleteAllLocalTaskVariables(java.lang.String,javax.servlet.http.HttpServletResponse)
07:29:40,146 [localhost-startStop-1] INFO  org.springframework.web.servlet.mvc.method.annotation.RequestMappingH
andlerMapping  - Mapped "{[/runtime/tasks/{taskId}/variables/{variableName}/data],methods=[GET],params=[],header
s=[],consumes=[],produces=[application/json],custom=[]}" onto public byte[] org.activiti.rest.service.api.runtim
e.task.TaskVariableDataResource.getVariableData(java.lang.String,java.lang.String,javax.servlet
.http.HttpServletRequest,javax.servlet.http.HttpServletResponse)
07:29:40,146 [localhost-startStop-1] INFO  org.springframework.web.servlet.mvc.method.annotation.RequestMappingH
andlerMapping  - Mapped "{[/runtime/tasks/{taskId}/variables/{variableName}],methods=[GET],params=[],headers=[],
consumes=[],produces=[application/json],custom=[]}" onto public org.activiti.rest.service.api.engine.variable.Re
stVariable org.activiti.rest.service.api.runtime.task.TaskVariableResource.getVariable(java.lang.String,java.lan
g.String,java.lang.String,javax.servlet.http.HttpServletRequest,javax.servlet.http.HttpServletResponse)
07:29:40,146 [localhost-startStop-1] INFO  org.springframework.web.servlet.mvc.method.annotation.RequestMappingH
andlerMapping  - Mapped "{[/runtime/tasks/{taskId}/variables/{variableName}],methods=[DELETE],params=[],headers=
[],consumes=[],produces=[],custom=[]}" onto public void org.activiti.rest.service.api.runtime.task.TaskVariableR
esource.deleteVariable(java.lang.String,java.lang.String,java.lang.String,javax.servlet.http.HttpServletResponse
)
07:29:40,146 [localhost-startStop-1] INFO  org.springframework.web.servlet.mvc.method.annotation.RequestMappingH
andlerMapping  - Mapped "{[/runtime/tasks/{taskId}/variables/{variableName}],methods=[PUT],params=[],headers=[],
consumes=[],produces=[application/json],custom=[]}" onto public org.activiti.rest.service.api.engine.variable.Re
stVariable org.activiti.rest.service.api.runtime.task.TaskVariableResource.updateVariable(java.lang.String,java.
lang.String,java.lang.String,javax.servlet.http.HttpServletRequest)
07:29:40,616 [localhost-startStop-1] INFO  org.springframework.web.servlet.mvc.method.annotation.RequestMappingH
andlerAdapter  - Looking for @ControllerAdvice: WebApplicationContext for namespace 'dispatcher-servlet': startu
p date [Wed Jul 27 19:29:39 PDT 2016]; parent: Root WebApplicationContext
07:29:40,653 [localhost-startStop-1] INFO  org.springframework.web.servlet.mvc.method.annotation.ExceptionHandle
rExceptionResolver  - Detected @ExceptionHandler methods in exceptionHandlerAdvice
07:29:40,686 [localhost-startStop-1] INFO  org.springframework.web.servlet.DispatcherServlet  - FrameworkServlet
'dispatcher': initialization completed in 1162 ms
Jul 27, 2016 7:29:40 PM org.apache.coyote.AbstractProtocol start
INFO: Starting ProtocolHandler ["http-bio-8080"]
```

Figure 8-7. *Activiti script running successfully*

A screen dump of the Activiti program running successfully is shown in Figure 8-7.

A picture of the Activiti Explorer dashboard being run successfully is shown in Figure 8-8.

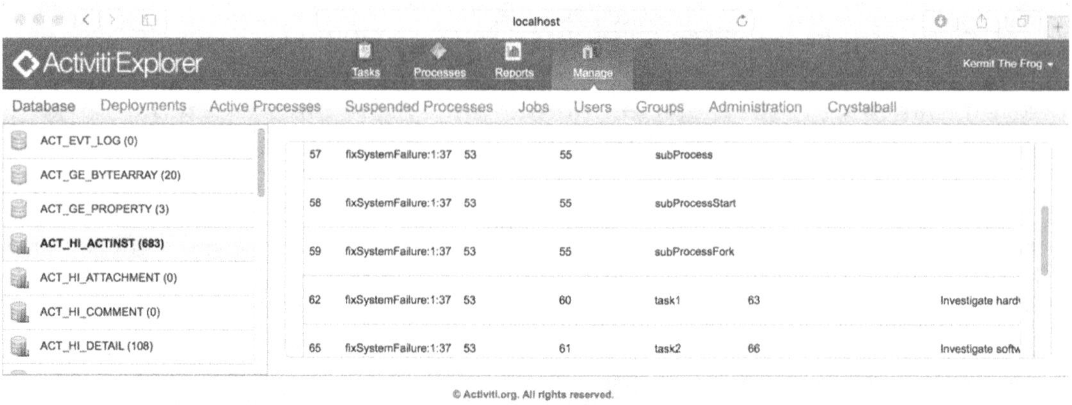

Figure 8-8. *Activiti explorer dashboard running successfully*

8.4   Analytical Engine Example with Rule Control

In this section, we will demonstrate an analytical engine example with rule control.

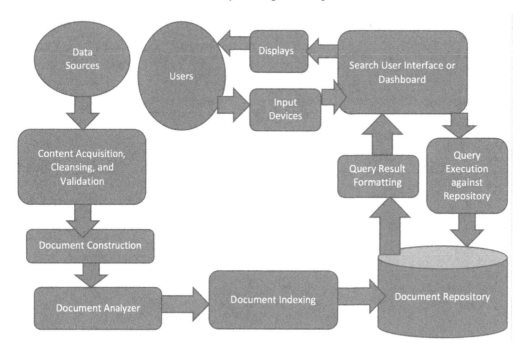

Figure 8-9. *An initial Lucene-oriented system design, including user interactions and document processing*

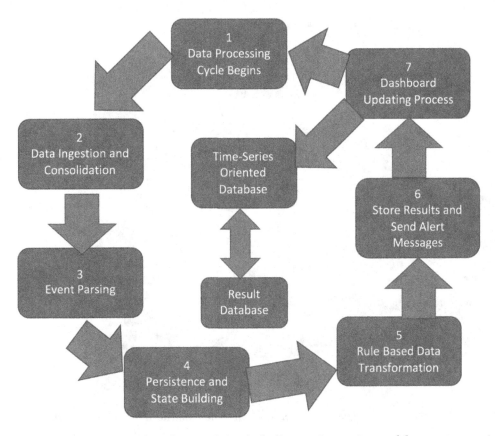

Figure 8-10. *A Lucene-oriented system design, including user interactions and document processing, step 2*

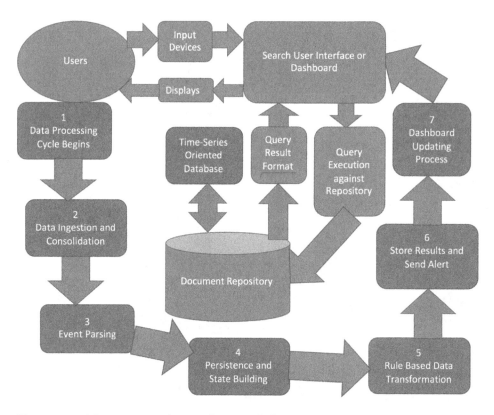

Figure 8-11. *A Lucene-oriented system design, including user interactions and document processing, step 3*

We can use the Splunk system for data ingestion.

We can use a time series–oriented database such as OpenTSDB (`https://github.com/OpenTSDB/opentsdb/releases`) as an intermediate data repository.

Rule-based transformation is provided by JBoss Drools.

The document repository functionality can be provided by an instance of a Cassandra database.

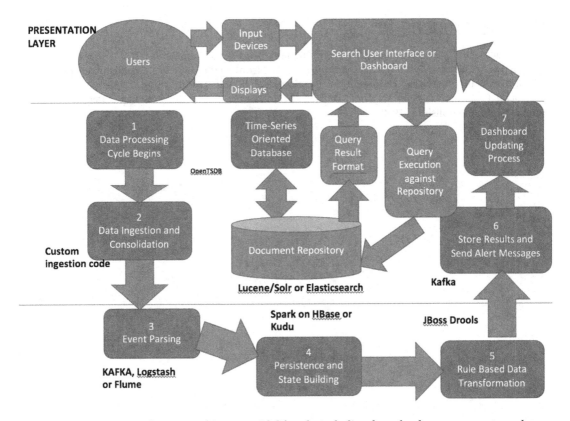

Figure 8-12. *An integrated system architecture with lifecycle, including the technology components used*

■ **Note**   Please note you are under no obligation use the technology components shown in Figure 8-12. You could use an alternative messaging component such as RabbitMQ, for example, instead of Apache Kafka, or MongoDB instead of Cassandra, depending upon your application requirements.

8.5 Summary

In this chapter, we discussed using rule-based controllers with other distributed components, especially with Hadoop and Spark ecosystem components. We have seen that a rule-based strategy can add a key ingredient to distributed data analytics: the ability to organize and control data flow in a flexible and logically organized manner. Scheduling and prioritization is a natural consequence of these rule-based techniques, and we looked at some examples of rule-based schedulers throughout the chapter.

In the next chapter, we will talk about using the techniques we have learned so far into one integrated analytical component which is applicable to a variety of use cases and problem domains.

8.6   References

Amador, Lucas. *Drools Developer Cookbook*. Birmingham, UK: PACKT Publishing, 2012.

Bali, Michal. *Drools JBoss Rules 5.0 Developers Guide*. Birmingham, UK: PACKT Publishing, 2009.

Browne, Paul. *JBoss Drools Business Rules*. Birmingham, UK: PACKT Publishing, 2009.

Norvig, Peter. *Paradigms of Artificial Intelligence: Case Studies in Common Lisp*. San Mateo, CA: Morgan-Kaufman Publishing, 1992.

CHAPTER 9

■ ■ ■

Putting It All Together: Designing a Complete Analytical System

In this chapter, we describe an end-to-end design example, using many of the components discussed so far. We also discuss "best practices" to use during the requirements acquisition, planning, architecting, development, testing, and deployment phases of the system development project.

■ **Note**   This chapter makes use of many of the software components discussed elsewhere throughout the book, including Hadoop, Spark, Splunk, Mahout, Spring Data, Spring XD, Samza, and Kafka. Check Appendix A for a summary of the components and insure that you have them available when trying out the examples from this chapter.

Building a complete distributed analytical system is easier than it sounds. We have already discussed many of the important ingredients for such a system in earlier chapters. Once you understand what your data sources and sinks are going to be, and you have a reasonably clear idea of the technology stack to be used and the "glueware" to be leveraged, writing the business logic and other processing code can become a relatively straightforward task.

A simple end-to-end architecture is shown in Figure 9-1. Many of the components shown allow some leeway as to what technology you actually use for data source, processors, data sinks and repositories, and output modules, which include the familiar dashboards, reports, visualizations, and the like that we will see in other chapters. In this example, we will use the familiar importing tool Splunk to provide an input source.

© Kerry Koitzsch 2017

K. Koitzsch, *Pro Hadoop Data Analytics*, DOI 10.1007/978-1-4842-1910-2_9

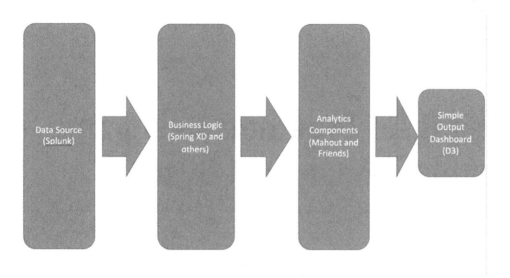

Figure 9-1. *A simple end-to-end analytics architecture*

In the following section we will describe how to set up and integrate Splunk with the other components of our example system.

HOW TO INSTALL SPLUNK FOR THE EXAMPLE SYSTEM

Splunk (https://www.splunk.com) is a logging framework and is very easy to download, install, and use. It comes with a number of very useful features for the kind of example analytics systems we're demonstrating here, including a built-in search facility.

To install Splunk, go to the download web page, create a user account, and download Splunk Enterprise for your appropriate platform. All the examples shown here are using the MacOS platform.

Install the Splunk Enterprise appropriately for your chosen platform. On the Mac platform, if the installation is successful, you will see Splunk represented in your Applications directories as shown in Figure 9-2.

Refer to http://docs.splunk.com/Documentation/Splunk/6.4.2/SearchTutorial/StartSplunk on how to start Splunk. Please note that the Splunk Web Interface can be found at http://localhost:8000 when started correctly.

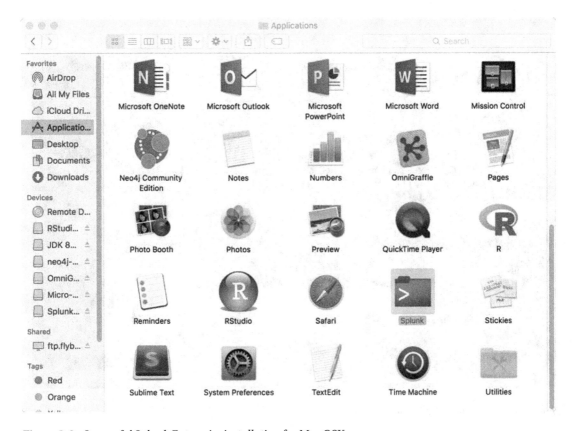

Figure 9-2. Successful Splunk Enterprise installation for Mac OSX

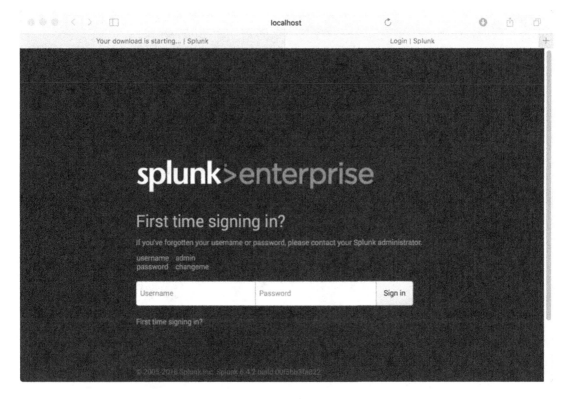

Figure 9-3. *Login page for Splunk Enterprise*

When you point your browser at localhost:8000, you'll initially see the Splunk login page. Use the default user name and password to begin with, change as instructed, and make sure the Java code you use for connectivity uses your updated username ('admin') and password('changename').

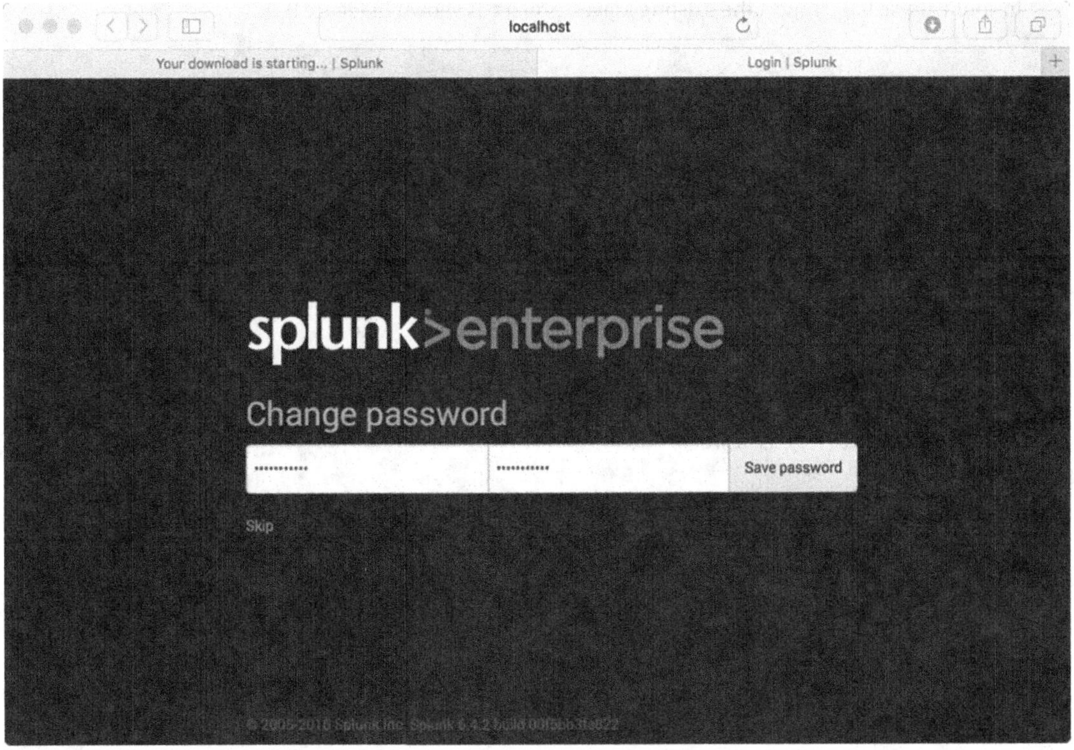

Figure 9-4. Change password during initial Splunk Enterprise setup

Download the following very useful library, splunk-library-javalogging, from github:

```
git clone https://github.com/splunk/splunk-library-javalogging.git

cd splunk-library-javalogging

mvn clean install
```

In your Eclipse IDE, import the existing Maven project as shown in Figure 9-5.

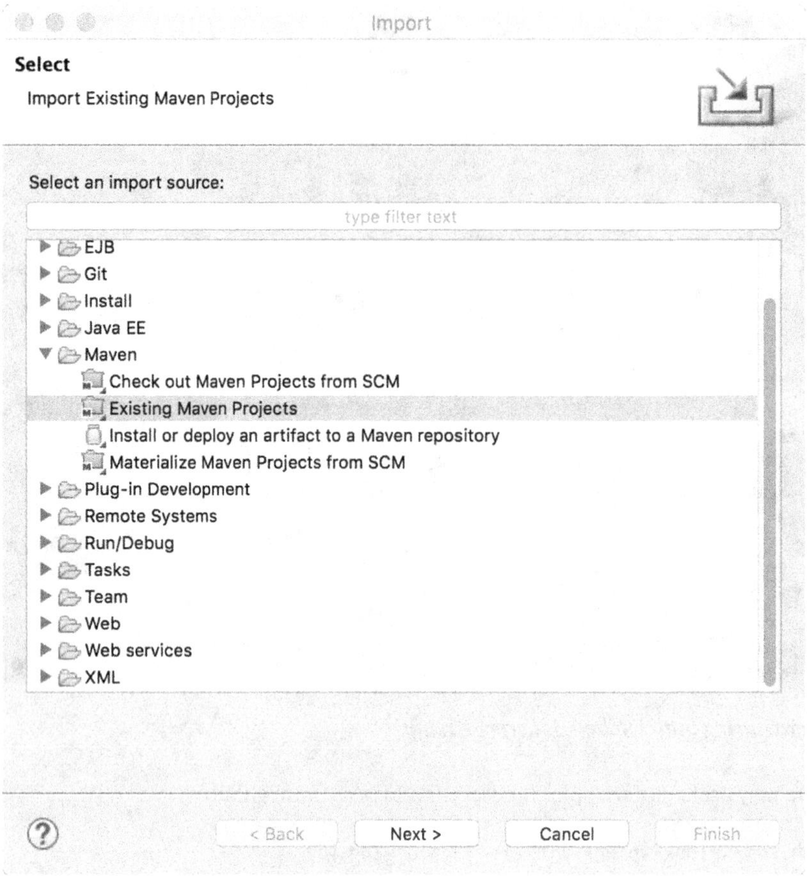

Figure 9-5. *Import an existing Maven to use splunk-library-javalogging*

Figure 9-5 shows a dialog for importing the existing Maven project to use splunk-library-javalogging.

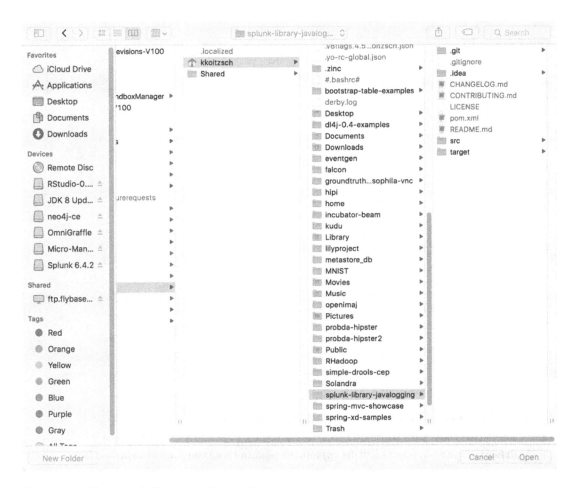

Figure 9-6. *Select splunk-library-javalogging for import*

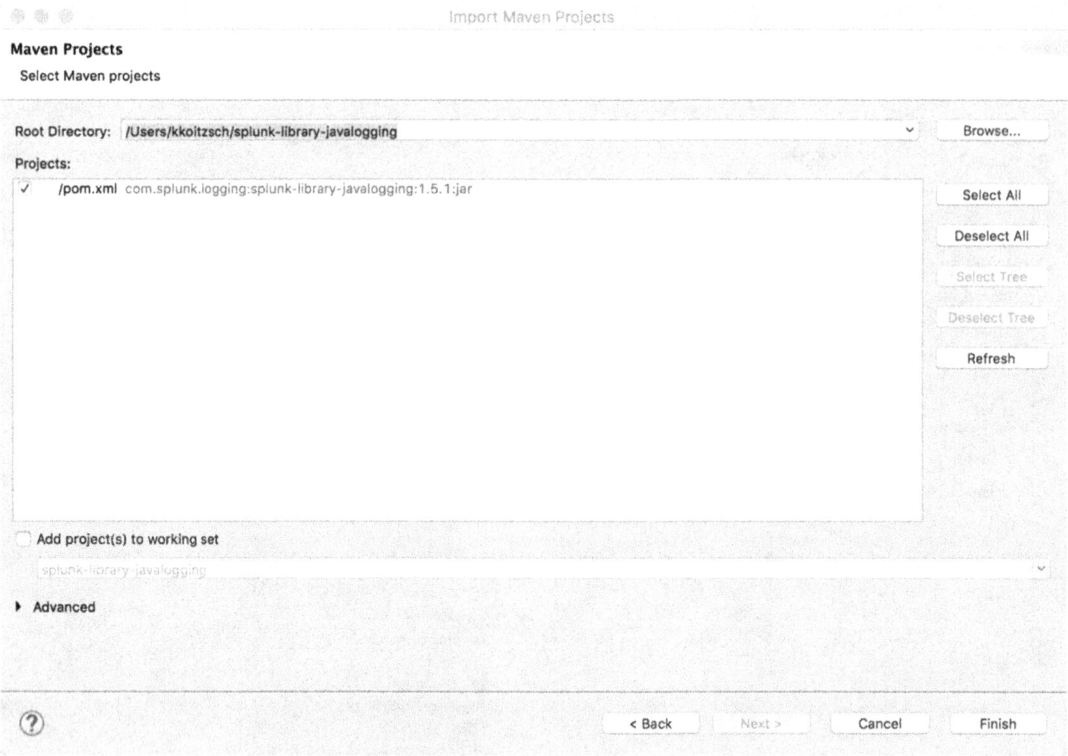

Figure 9-7. *Select the appropriate root directory for Maven construction*

As shown in Figure 9-7, selection of the appropriate pom.xml is all you need to do in this step.

Figure 9-8. *Eclipse IDE installation of Splunk test code*

As shown in Figure 9-8, modification to include appropriate username and password values is typically all that is necessary for this step of installation.

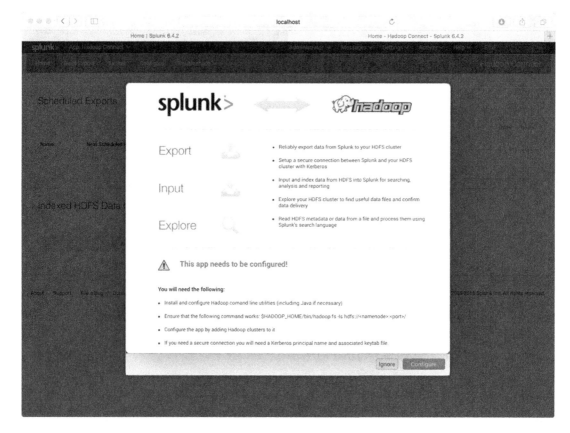

Figure 9-9. *Configure the HadoopConnect component for Splunk*

Configure the HadoopConnect component for Splunk as shown in Figure 9-9.

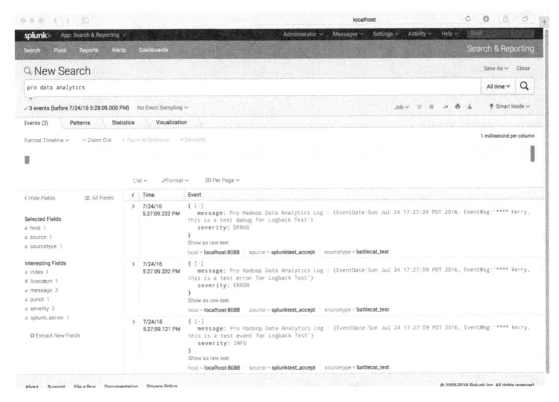

Figure 9-10. *Searching for Pro Data Analytics events in the Splunk dashboard*

Textual search in the Splunk dashboard can be accomplished as in Figure 9-10. We can also select an appropriate timestamped interval to perform queries over our data set.

Visualization is an important part of this integration process. Check out some of the D3 references at the bottom of this chapter to get a sense of some of the techniques you can use in combination with the other components of the data pipeline.

9.1   Summary

In this chapter, we discussed building a complete analytical system and some of the challenges architects and developers encounter upon the way. We constructed a complete end-to-end analytics pipeline using the now-familiar technology components discussed in earlier chapters. In particular, we talked about how to use Splunk as an input data source. Splunk is a particularly versatile and flexible tool for all kinds of generic logging events.

9.2   References

Mock, Derek, Johnson, Paul R., Diakun, Josh. Splunk Operational Intelligence Cookbook. Birmingham, UK: PACKT Publishing, 2014.

Zhu, Nick Qi. Data Visualization with d3.js Cookbook. Birmingham, UK: PACKT Publishing, 2014.

PART III

Components and Systems

The third part of our book describes the component parts and associated libraries which can assist us in building distributed analytic systems. This includes components based on a variety of different programming languages, architectures, and data models.

■ ■ ■

CHAPTER 10

Data Visualizers: Seeing and Interacting with the Analysis

In this chapter, we will talk about how to look at—to visualize—our analytical results. This is actually quite a complex process, or it can be. It's all a matter of choosing an appropriate technology stack for the kind of visualizing you need to do for your application. The visualization task in an analytics application can range from creating simple reports to full-fledged interactive systems. In this chapter we will primarily be discussing Angular JS and its ecosystem, including the ElasticUI visualization tool Kibana, as well as other visualization components for graphs, charts, and tables, including some JavaScript-based tools like D3.js and sigma.js.

10.1   Simple Visualizations

One of the simplest visualization architectures is shown in Figure 10-1. The front-end control interface may be web-based, or a stand-alone application. The control UI may be based on a single web page, or a more developed software plug-in or multiple page components. "Glueware" on the front end might involve visualization frameworks such as Angular JS, which we will discuss in detail in the following sections. On the back end, glueware such as Spring XD can make interfacing to a visualizer much simpler.

© Kerry Koitzsch 2017
K. Koitzsch, *Pro Hadoop Data Analytics*, DOI 10.1007/978-1-4842-1910-2_10

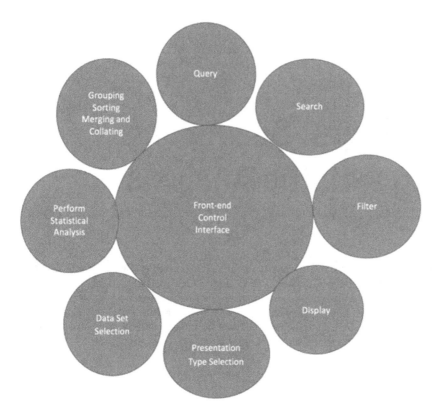

Figure 10-1. *Typical visualization component architecture*

Let's talk briefly about the different components in Figure 10-1. Each circle represents different facets of typical use cases when using an analytics software component. You might think of the circles as individual sub-problems or issues we are trying to solve. For example, grouping, sorting, merging, and collating might be handled by a standard tabular structure, such as the one shown in Figure 10-2. Most of the sorting and grouping problems are solved with built-in table functionality like clicking on a column to sort rows, or to group items.

Providing effective display capabilities can be as simple as selecting an appropriate tabular component to use for row-oriented data. A good example of a tabular component which provides data import, sorting, pagination, and easily programmable features is the one shown in Figure 10-2. This component is available at `https://github.com/wenzhixin/bootstrap-table`. The control shown here leverages a helper library called Bootstrap.js (`http://getbootstrap.com/javascript/`) to provide the advanced functionality. Being able to import JSON data sets into a visualization component is a key feature which enables seamless integration with other UI and back-end components.

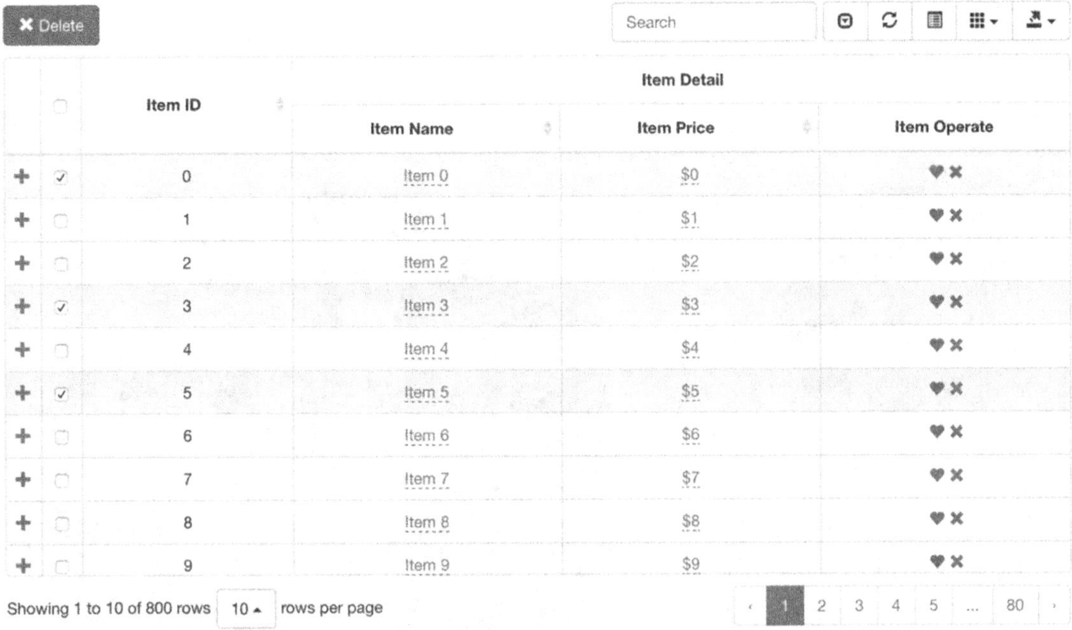

✖ Delete			Search 🕓 🔁 ▤ ▦ ▾ 🏷 ▾		

	☐	Item ID ⇕	Item Detail		
			Item Name ⇕	Item Price ⇕	Item Operate
✚	☑	0	Item 0	$0	♥ ✖
✚	☐	1	Item 1	$1	♥ ✖
✚	☐	2	Item 2	$2	♥ ✖
✚	☑	3	Item 3	$3	♥ ✖
✚	☐	4	Item 4	$4	♥ ✖
✚	☑	5	Item 5	$5	♥ ✖
✚	☐	6	Item 6	$6	♥ ✖
✚	☐	7	Item 7	$7	♥ ✖
✚	☐	8	Item 8	$8	♥ ✖
✚	☐	9	Item 9	$9	♥ ✖

Showing 1 to 10 of 800 rows ‹ 10 ▲ › rows per page   ‹ **1** 2 3 4 5 ... 80 ›

Figure 10-2. *One tabular control can solve several visualization concerns*

Many of the concerns found in Figure 10-1 can be controlled by front-end controls we embed in a web page. For example, we are all familiar with the "Google-style" text search mechanism, which consists of just a text field and a button. We can implement a visualization tool using d3 that does simple analytics on Facebook tweets as an introduction to data visualization. As shown in Figure 10-2 and Figure 10-3, we can control the "what" of the display as well as the "how": we can see a pie chart, bar chart, and bubble chart version of the sample data set, which is coming from a Spring XD data stream.

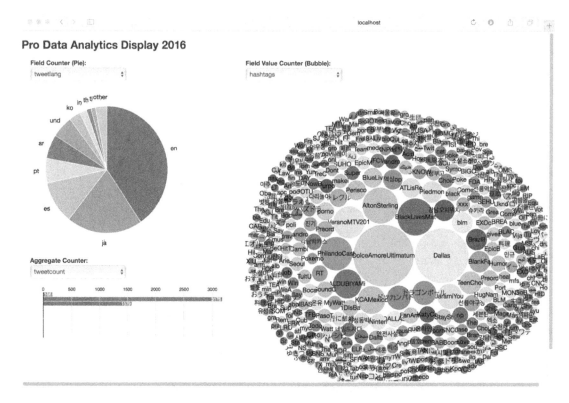

Figure 10-3. *Simple data visualization example of Twitter tweets using Spring XD showing trending topics and languages*

Most of the concerns we see in Figure 10-1 (data set selection, presentation type selection, and the rest) are represented in Figure 10-3 and Figure 10-4. Standard controls, such as drop-down boxes, are used to select data sets and presentation types. Presentation types may include a wide range of graph and chart types, two- and three-dimensional display, and other types of presentation and report formats. Components such as Apache POI (`https://poi.apache.org`) may be used to write report files in Microsoft formats compatible with Excel.

The display shown here dynamically updates as new tweet data arrives through the Spring XD data streams. Figure 10-3 shows a slightly different visualization of the tweet data, in which we can see how some circles grow in size, representing the data "trending" in Twitter.

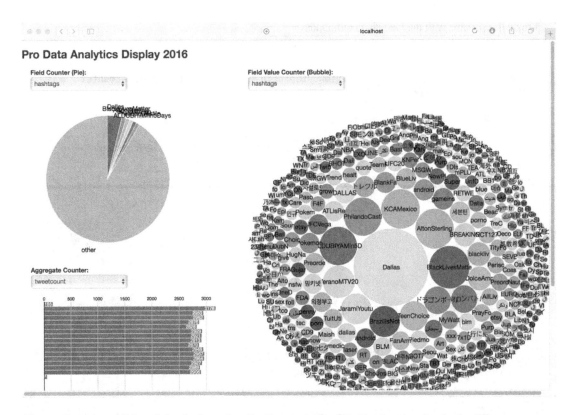

Figure 10-4. An additional simple data visualization example of Twitter tweets using Spring XD

We'll discuss Spring XD in the next section, as it is particularly useful as glueware when building visualizers.

SETTING UP THE SPRING XD COMPONENT

Setting up the Spring XD component, like all the Spring Framework components, is basically straightforward.

After installing Spring XD, start Spring XD in "single node mode" with

```
bin/xd-singlenode
cd bin
```

Run the XD shell with the command

```
./xd-shell
```

Create the streams with the following commands

```
stream create tweets --definition "twitterstream | log"

stream create tweetlang  --definition "tap:stream:tweets > field-value-counter
--fieldName=lang" --deploy

stream create tweetcount --definition "tap:stream:tweets > aggregate-counter" --deploy

stream create tagcount --definition "tap:stream:tweets > field-value-counter
--fieldName=entities.hashtags.text --name=hashtags" --deploy
stream deploy tweets
```

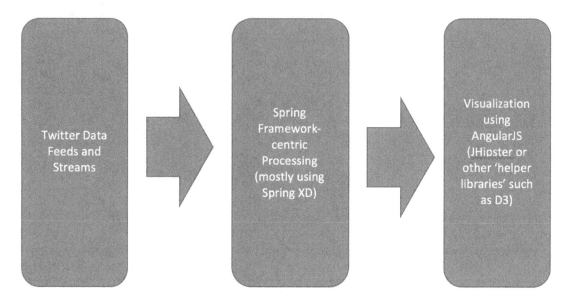

Figure 10-5. *Architecture diagram for Twitter ➤ Spring XD ➤ visualization*

```
● ● ●          ▦ bin — XD Shell 1.3.1.RELEASE — java — 80×24
// XD Shell 1.1.0.RELEASE log opened at 2016-05-19 16:11:20           ▣
stream create --definition "time | log" --name ticktock --deploy
quit
// XD Shell 1.1.0.RELEASE log closed at 2016-05-19 16:12:17
// XD Shell 1.1.0.RELEASE log opened at 2016-05-19 16:13:42
stream create --definition "time | log" --name pro-big-data-analytics --deploy
// XD Shell 1.1.0.RELEASE log opened at 2016-07-07 13:01:14
help
quit
// XD Shell 1.1.0.RELEASE log closed at 2016-07-07 13:01:51
Kerrys-MacBook-Pro-2:bin kerryk$ ./xd-shell
Unable to find a $JAVA_HOME at "/usr", continuing with system-provided Java...

  /￣￣|        (-)         \ \ / / _  ￣￣\
  \ `=._ _    _    _  _  _   \ V /| | | |
   `--.\|`=\| |`=\|/ |_ | /^\| | | |
  /\_/ / |_) | | | | | (_| | / / \ \ |/ /
  \___/| .__/|_| |_|_| |_|\_, | \/   \/\__/
       | |               _/ |
       |_|              |__/
eXtreme Data
1.3.1.RELEASE | Admin Server Target: http://localhost:9393
Welcome to the Spring XD shell. For assistance hit TAB or type "help".
xd:>▯
```

Figure 10-6. *Bringing up the Spring XD shell successfully*

Figure 10-7. *Using Spring XD to implement a Twitter tweet stream and then sdeploy the stream*

In the next section we will go into some comprehensive examples of a particularly useful toolkit, Angular JS.

10.2   Introducing Angular JS and Friends

Angular JS (https://angularjs.org) is a JavaScript-based toolkit that has become a very prominent contender in the data visualization library arena. It has a straightforward model-view-controller (MVC) architecture which enables a streamlined design and implementation process.

Incidentally, some Angular JS components such as Elastic UI (elasticui.com) are available directly out of the box to use with the Elastic search engine. ElasticUI with Kibana is a quck and relatively painless way to add visualization components.

We will spend most of the rest of this chapter discussing how to set up some examples using Angular JS and some other visualization toolkits, including a very interesting new arrival on the scene, JHipster.

10.3   Using JHipster to Integrate Spring XD and Angular JS

JHipster (https://jhipster.github.io) is an open source Yeoman ( ) generator designed to create integrated Spring Boot and Angular JS components. This makes it possible to integrate additional components from the rest of the Spring Framework ecosystem as well in a seamless manner. For example, you could use a Spring Data Hadoop-based component to build a data pipeline with summary displays written in AngularJS on the front end.

We are going to build a simple JHipster mini-project to show how this might work.

Figure 10-8. *Successful setup of a "probda-hipster" project*

HOW TO BUILD THE ANGULAR JS EXAMPLE SYSTEM

Building an Angular JS example system is relatively straightforward and we describe how to do it in this section.

The first step in building the Angular JS example system is to make the archetype project on the command line. Cd to the home directory you wish to build in. Then execute the following command, as shown in Listing 13.1.

```
mvn archetype:generate -DgroupId=nl.ivonet -DartifactId=java-angularjs-seed
-DarchetypeArtifactId=maven-archetype-webapp -DinteractiveMode=false
```

This will create the directories and files shown in Listing 10-2. Cd to the directory and make sure they are really there.

```
./pom.xml
./src
./src/main
./src/main/resources
./src/main/webapp
./src/main/webapp/index.jsp
./src/main/webapp/WEB-INF
./src/main/webapp/WEB-INF/web.xml
```

Construct the new files and directories to configure the project, as shown in Listing 10-3.

```
mkdir -p src/main/java
mkdir -p src/test/java
mkdir -p src/test/javascript/unit
mkdir -p src/test/javascript/e2e
mkdir -p src/test/resources
rm -f ./src/main/webapp/WEB-INF/web.xml
rm -f ./src/main/webapp/index.jsp
mkdir -p ./src/main/webapp/css
touch ./src/main/webapp/css/specific.css
mkdir -p ./src/main/webapp/js
touch ./src/main/webapp/js/app.js
touch ./src/main/webapp/js/controllers.js
touch ./src/main/webapp/js/routes.js
touch ./src/main/webapp/js/services.js
touch ./src/main/webapp/js/filters.js
touch ./src/main/webapp/js/services.js
mkdir -p ./src/main/webapp/vendor
mkdir -p ./src/main/webapp/partials
mkdir -p ./src/main/webapp/img
touch README.md
touch .bowerrc
```

Run the npm initialization to interactively build the program. 'npm init' will provide a step-by-step question-and-answer approach towards creating the project, as shown in Listing x.y.

```
npm init

This utility will walk you through creating a package.json file.
It only covers the most common items, and tries to guess sane defaults.

See `npm help json` for definitive documentation on these fields
and exactly what they do.

Use `npm install  --save` afterwards to install a package and
save it as a dependency in the package.json file.

Press ^C at any time to quit.
name: (java-angularjs-seed)
version: (0.0.0)
description: A starter project for AngularJS combined with java and maven
entry point: (index.js)
test command: karma start test/resources/karma.conf.js
git repository: https://github.com/ivonet/java-angular-seed
keywords:
author: Ivo Woltring
license: (ISC) Apache 2.0
About to write to /Users/ivonet/dev/ordina/LabTime/java-angularjs-seed/package.json:
```

```
{
  "name": "java-angularjs-seed",
  "version": "0.0.0",
  "description": "A starter project for AngularJS combined with java and maven",
  "main": "index.js",
  "scripts": {
    "test": "karma start test/resources/karma.conf.js"
  },
  "repository": {
    "type": "git",
    "url": "https://github.com/ivonet/java-angular-seed"
  },
  "author": "Ivo Woltring",
  "license": "Apache 2.0",
  "bugs": {
    "url": "https://github.com/ivonet/java-angular-seed/issues"
  },
  "homepage": "https://github.com/ivonet/java-angular-seed"
}
```

Is this ok? (yes)

Now add the following content to the file: _____.

```
{
  "name": "java-angular-seed",
  "private": true,
  "version": "0.0.0",
  "description": "A starter project for AngularJS combined with java and maven",
  "repository": "https://github.com/ivonet/java-angular-seed",
  "license": "Apache 2.0",
  "devDependencies": {
    "bower": "^1.3.1",
    "http-server": "^0.6.1",
    "karma": "~0.12",
    "karma-chrome-launcher": "^0.1.4",
    "karma-firefox-launcher": "^0.1.3",
    "karma-jasmine": "^0.1.5",
    "karma-junit-reporter": "^0.2.2",
    "protractor": "~0.20.1",
    "shelljs": "^0.2.6"
  },
  "scripts": {
    "postinstall": "bower install",
    "prestart": "npm install",
    "start": "http-server src/main/webapp -a localhost -p 8000",
    "pretest": "npm install",
    "test": "karma start src/test/javascript/karma.conf.js",
    "test-single-run": "karma start src/test/javascript/karma.conf.js  --single-run",
    "preupdate-webdriver": "npm install",
    "update-webdriver": "webdriver-manager update",
```

```
    "preprotractor": "npm run update-webdriver",
    "protractor": "protractor src/test/javascript/protractor-conf.js",
    "update-index-async": "node -e \"require('shelljs/global'); sed('-i', /\\/\\/@@NG_
LOADER_START@@[\\s\\S]*\\/\\/@@NG_LOADER_END@@/, '//@@NG_LOADER_START@@\\n' + cat('src/
main/webapp/vendor/angular-loader/angular-loader.min.js') + '\\n//@@NG_LOADER_END@@',
'src/main/webapp/index.html');\""
  }
}
```

```
                              🏠 kerryk — bash — 112×34
Kerrys-MacBook-Pro:demo kerryk$ cd
Kerrys-MacBook-Pro:~ kerryk$ mvn archetype:generate -DgroupId=com.apress -DartifactId=probda -DarchetypeArtifact
Id=maven-archetype-webapp -DinteractiveMode=false
Unable to find a $JAVA_HOME at "/usr", continuing with system-provided Java...
[INFO] Scanning for projects...
[INFO]
[INFO] ------------------------------------------------------------------------
[INFO] Building Maven Stub Project (No POM) 1
[INFO] ------------------------------------------------------------------------
[INFO]
[INFO] >>> maven-archetype-plugin:2.4:generate (default-cli) > generate-sources @ standalone-pom >>>
[INFO]
[INFO] <<< maven-archetype-plugin:2.4:generate (default-cli) < generate-sources @ standalone-pom <<<
[INFO]
[INFO] --- maven-archetype-plugin:2.4:generate (default-cli) @ standalone-pom ---
[INFO] Generating project in Batch mode
[INFO] ------------------------------------------------------------------------
[INFO] Using following parameters for creating project from Old (1.x) Archetype: maven-archetype-webapp:1.0
[INFO] ------------------------------------------------------------------------
[INFO] Parameter: basedir, Value: /Users/kerryk
[INFO] Parameter: package, Value: com.apress
[INFO] Parameter: groupId, Value: com.apress
[INFO] Parameter: artifactId, Value: probda
[INFO] Parameter: packageName, Value: com.apress
[INFO] Parameter: version, Value: 1.0-SNAPSHOT
[INFO] project created from Old (1.x) Archetype in dir: /Users/kerryk/probda
[INFO] ------------------------------------------------------------------------
[INFO] BUILD SUCCESS
[INFO] ------------------------------------------------------------------------
[INFO] Total time: 4.637 s
[INFO] Finished at: 2016-04-26T16:34:31-07:00
[INFO] Final Memory: 12M/189M
[INFO] ------------------------------------------------------------------------
Kerrys-MacBook-Pro:~ kerryk$ ▯
```

Figure 10-9. *Building the Maven stub for the Angular JS project successfully on the command line*

```
                              probda — bash — 112×34
? would you like to mark this package as private which prevents it from being accidentally published to the regi
stry? Yes
stry? (y/N) y
{try? (y/N)
  name: 'probda',
  description: 'A starter project for AngularJS combined with java and maven',
  main: 'src/main/webapp/index.html',
  authors: [
    'Kerry Koitzsch <kkoitzsch@apple.com>'
  ],
  license: 'Apache 2.0',
  moduleType: [],
  homepage: 'http://probda.html',
  private: true,
  ignore: [
    '**/.*',
    'node_modules',
    'bower_components',
    'src/main/webapp/probda',
    'test',
    'tests'
  ],
  dependencies: {
    'angular-loader': '1.3.0-beta.14',
    'angular-route': '1.3.0-beta.14',
    'angular-mocks': '1.3.0-beta.14',
    'angular-animate': '1.3.0-beta.14',
    bootstrap: '^3.3.6',
    angular: '1.3.0-beta.14'
  }
}

? Looks good? Yes
Kerrys-MacBook-Pro:probda kerryk$ []
```

Figure 10-10. *Configuration file for the Angular JS example*

```
                              probda — node — 112×34
Kerrys-MacBook-Pro:probda kerryk$ npm init
This utility will walk you through creating a package.json file.
It only covers the most common items, and tries to guess sensible defaults.

See `npm help json` for definitive documentation on these fields
and exactly what they do.

Use `npm install <pkg> --save` afterwards to install a package and
save it as a dependency in the package.json file.

Press ^C at any time to quit.
name: (probda)
version: (1.0.0)
description: basic Pro Hadoop Analytics project with AngularJS
entry point: (index.js)
test command: karma start test/resources/karma.conf.js
git repository:
keywords:
author: Kerry Koitzsch
license: (ISC)
About to write to /Users/kerryk/probda/package.json:

{
  "name": "probda",
  "version": "1.0.0",
  "description": "basic Pro Hadoop Analytics project with AngularJS",
  "main": "index.js",
  "scripts": {
    "test": "karma start test/resources/karma.conf.js"
  },
  "author": "Kerry Koitzsch",
  "license": "ISC"
}
```

Figure 10-11. *Additional configuration file for the Angular JS example application*

```
{
    "directory": "src/main/webapp/vendor"
}
bower install angular#1.3.0-beta.14
bower install angular-route#1.3.0-beta.14
bower install angular-animate#1.3.0-beta.14
bower install angular-mocks#1.3.0-beta.14
bower install angular-loader#1.3.0-beta.14
bower install bootstrap

bower init
[?] name: java-angularjs-seed
[?] version: 0.0.0
[?] description: A java / maven / angularjs seed project
[?] main file: src/main/webapp/index.html
[?] what types of modules does this package expose?
[?] keywords: java,maven,angularjs,seed
[?] authors: IvoNet
[?] license: Apache 2.0
[?] homepage: http://ivonet.nl
[?] set currently installed components as dependencies? Yes
[?] add commonly ignored files to ignore list? Yes
[?] would you like to mark this package as private which prevents it from being
accidentally pub[?] would you like to mark this package as private which prevents it
from being accidentally published to the registry? Yes


...


[?] Looks good? (Y/n) Y

{
    "name": "java-angularjs-seed",
    "version": "0.0.0",
    "authors": [
        "IvoNet <webmaster@ivonet.nl>"
    ],
    "description": "A java / maven / angularjs seed project",
    "keywords": [
        "java",
        "maven",
        "angularjs",
        "seed"
    ],
    "license": "Apache 2.0",
    "homepage": "http://ivonet.nl",
    "private": true,
    "ignore": [
        "**/.*",
        "node_modules",
        "bower_components",
        "src/main/webapp/vendor",
```

```
        "test",
        "tests"
    ],
    "dependencies": {
        "angular": "1.3.0-beta.14",
        "angular-loader": "1.3.0-beta.14",
        "angular-mocks": "1.3.0-beta.14",
        "angular-route": "1.3.0-beta.14",
        "bootstrap": "3.2.0"
    },
    "main": "src/main/webapp/index.html"
}
rm -rf ./src/main/webapp/vendor
npm install
```

Now we configure ./src/test/javascript/karma.conf.js :

```
module.exports = function(config){
  config.set({

    basePath : '../../../',

    files : [
      'src/main/webapp/vendor/angular**/**.min.js',
      'src/main/webapp/vendor/angular-mocks/angular-mocks.js',
      'src/main/webapp/js/**/*.js',
      'src/test/javascript/unit/**/*.js'
    ],

    autoWatch : true,

    frameworks: ['jasmine'],

    browsers : ['Chrome'],

    plugins : [
            'karma-chrome-launcher',
            'karma-firefox-launcher',
            'karma-jasmine',
            'karma-junit-reporter'
            ],

    junitReporter : {
      outputFile: 'target/test_out/unit.xml',
      suite: 'src/test/javascript/unit'
    }

  });
};
```

Figure 10-12. *Console result of Angular component install*

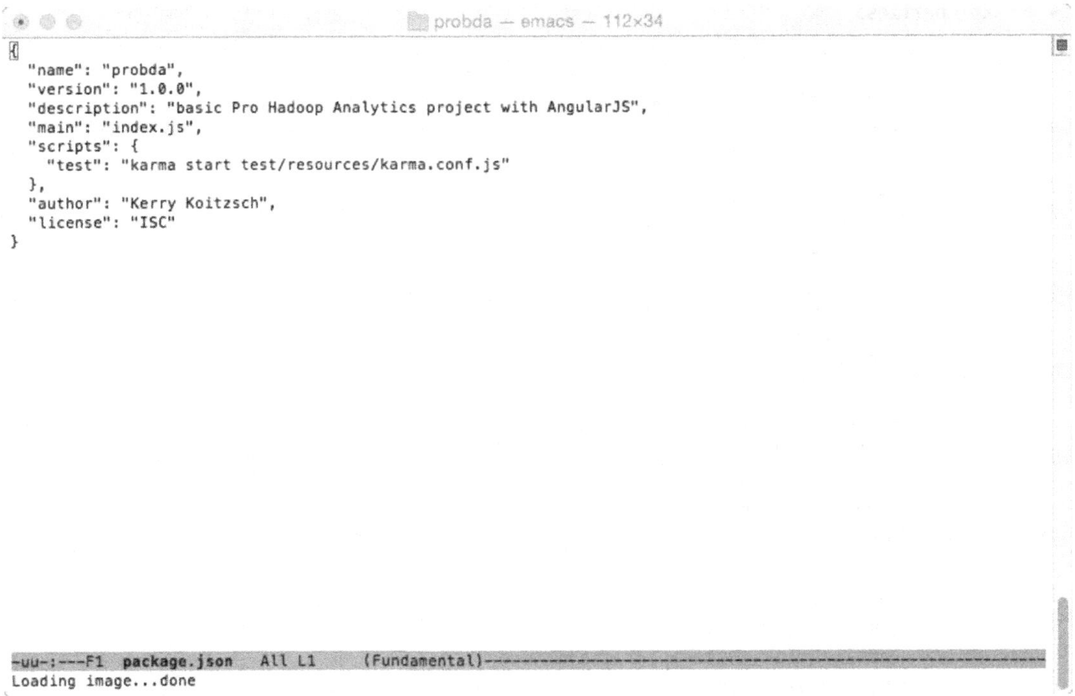

```
●  ●  ●                                    📁 probda — emacs — 112×34

🗌
 "name": "probda",
 "version": "1.0.0",
 "description": "basic Pro Hadoop Analytics project with AngularJS",
 "main": "index.js",
 "scripts": {
   "test": "karma start test/resources/karma.conf.js"
 },
 "author": "Kerry Koitzsch",
 "license": "ISC"
}
```

```
-uu-:---F1  package.json   All L1     (Fundamental)---------------------------------------------------------------
Loading image...done
```

Figure 10-13. *Data configuration in the package.json file*

Put the following content in ./src/main/webapp/WEB-INF/beans.xml:

```xml
<?xml version="1.0" encoding="UTF-8"?>
<beans xmlns="http://xmlns.jcp.org/xml/ns/javaee"
       xmlns:xsi="http://www.w3.org/2001/XMLSchema-instance"
       xsi:schemaLocation="http://xmlns.jcp.org/xml/ns/javaee http://xmlns.jcp.org/xml/
ns/javaee/beans_1_1.xsd"
       bean-discovery-mode="annotated">
</beans>
```

```xml
<project xmlns="http://maven.apache.org/POM/4.0.0" xmlns:xsi="http://www.w3.org/2001/
XMLSchema-instance"
         xsi:schemaLocation="http://maven.apache.org/POM/4.0.0 http://maven.apache.org/
maven-v4_0_0.xsd">
    <modelVersion>4.0.0</modelVersion>
    <groupId>nl.ivonet</groupId>
    <artifactId>java-angularjs-seed</artifactId>
    <packaging>war</packaging>
    <version>1.0-SNAPSHOT</version>

    <name>java-angularjs-seed Maven Webapp</name>

    <url>http://ivonet.nl</url>
```

```xml
<properties>
    <artifact.name>app</artifact.name>
    <endorsed.dir>${project.build.directory}/endorsed</endorsed.dir>
    <project.build.sourceEncoding>UTF-8</project.build.sourceEncoding>
</properties>

<dependencies>
    <dependency>
        <groupId>junit</groupId>
        <artifactId>junit</artifactId>
        <version>4.11</version>
        <scope>test</scope>
    </dependency>
    <dependency>
        <groupId>org.mockito</groupId>
        <artifactId>mockito-all</artifactId>
        <version>1.9.5</version>
        <scope>test</scope>
    </dependency>

    <dependency>
        <groupId>javax</groupId>
        <artifactId>javaee-api</artifactId>
        <version>7.0</version>
        <scope>provided</scope>
    </dependency>

</dependencies>
<build>
    <finalName>${artifact.name}</finalName>
    <plugins>
        <plugin>
            <groupId>org.apache.maven.plugins</groupId>
            <artifactId>maven-compiler-plugin</artifactId>
            <version>3.1</version>
            <configuration>
                <source>1.8</source>
                <target>1.8</target>
                <compilerArguments>
                    <endorseddirs>${endorsed.dir}</endorseddirs>
                </compilerArguments>
            </configuration>
        </plugin>
        <plugin>
            <groupId>org.apache.maven.plugins</groupId>
            <artifactId>maven-war-plugin</artifactId>
            <version>2.4</version>
            <configuration>
                <failOnMissingWebXml>false</failOnMissingWebXml>
            </configuration>
        </plugin>
```

```xml
<plugin>
    <groupId>org.apache.maven.plugins</groupId>
    <artifactId>maven-dependency-plugin</artifactId>
    <version>2.6</version>
    <executions>
        <execution>
            <phase>validate</phase>
            <goals>
                <goal>copy</goal>
            </goals>
            <configuration>
                <outputDirectory>${endorsed.dir}</outputDirectory>
                <silent>true</silent>
                <artifactItems>
                    <artifactItem>
                        <groupId>javax</groupId>
                        <artifactId>javaee-endorsed-api</artifactId>
                        <version>7.0</version>
                        <type>jar</type>
                    </artifactItem>
                </artifactItems>
            </configuration>
        </execution>
    </executions>
</plugin>
</plugins>
</build>
</project>
```

10.4   Using d3.js, sigma.js and Others

D3.js (https://d3js.org) and sigma.js (http://sigmajs.org) are popular JavaScript libraries for data visualization.

Examples of the graph visualizations which are possible with d3 and sigmajs toolkits

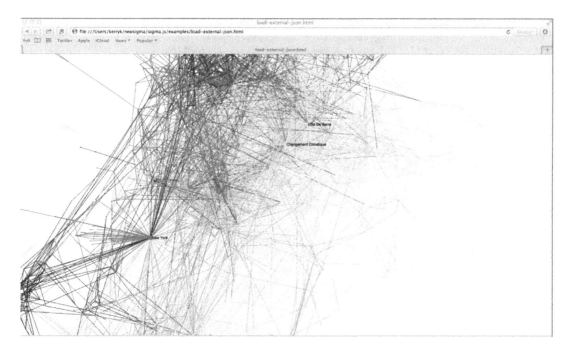

Figure 10-14. *A portion of a sigma.js-based graph visualization example*

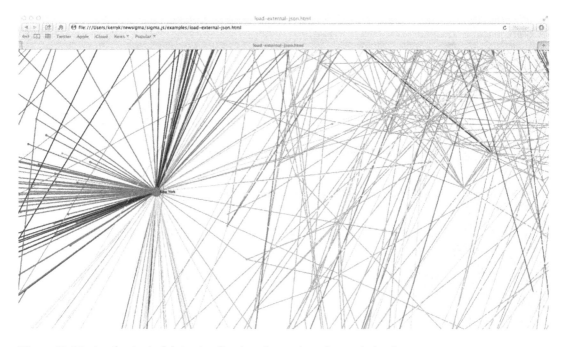

Figure 10-15. *Another typical data visualization of a portion of a graph database*

We can handcraft user interfaces to suit our application, or we have the option to use some of the sophisticated visualization tools already available as stand-alone libraries, plug-ins, and toolkits.

Recall that we can visualize data sets directly from graph databases as well. For example, in Neo4j, we can browse through the crime statistics of Sacramento after loading the CSV data set. Clicking on the individual nodes causes a summary of the fields to appear at the bottom of the graph display, as shown in Figure 10-16.

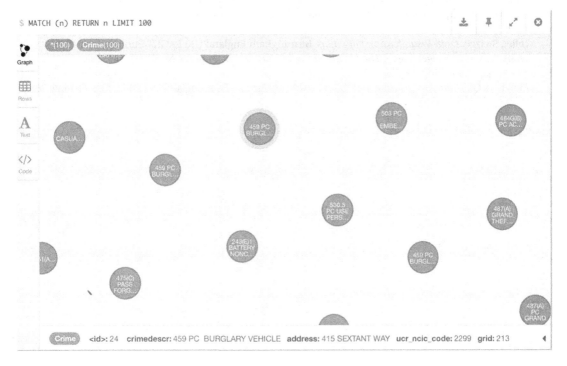

Figure 10-16. *Browsing crime statistics as individual nodes from a query in a Neo4j graph database*

10.5 Summary

In this chapter we looked at the visual side of the analytics problem: how to see and understand the results of our analytical processes. The solution to the visualization challenge can be as simple as a CSV report in Excel all the way up to a sophisticated interactive dashboard. We emphasized the use of Angular JS, a sophisticated visualization toolkit based on the model-view-controller (MVC) paradigm.

In the next chapter, we discuss rule-based control and orchestration module design and implementation. Rule systems are a type of control system with a venerable history in computer software, and have proven their effectiveness in a wide range of control and scheduling applications over the course of time.

We will discover that rule-based modules can be a useful component in distributed analytics systems, especially for scheduling and orchestrating individual processes within the overall application execution.

10.6   References

Ford, Brian, and Ruebbelke, Lukas. *Angular JS in Action*. Boston, MA: O'Reilly Publishing, 2015.

Freeman, Adam. *Pro AngularJS*. New York, NY: Apress Publishing, 2014.

Frisbie, Matt. *AngularJS Web Application Development Cookbook*. Birmingham England UK: PACKT Publishing, 2013.

Murray, Scott. *Interactive Data Visualization for the Web*. Boston, MA: O'Reilly Publishing, 2013.

Pickover, Clifford A., Tewksbury, Stuart K. (eds). *Frontiers of Scientific Visualization*. New York, NY: Wiley-Interscience, 1994,

Teller, Swizec. *Data Visualization with d3.js*. Birmingham England UK: PACKT Publishing 2013.

Wolff, Robert S., Yaeger, Larry. *The Visualization of Natural Phenomena*. New York, NY: Telos/Springer-Verlag Publishing, 1993.

Zhu, Nick Qi. *Data Visualization with D3.js Cookbook*. Birmingham England UK: PACKT Publishing, 2013.

PART IV

Case Studies and Applications

In the final part of our book, we examine case studies and applications of the kind of distributed systems we have discussed. We end the book with some thoughts about the future of Hadoop and distributed analytic systems in general.

CHAPTER 11

■ ■ ■

A Case Study in Bioinformatics: Analyzing Microscope Slide Data

In this chapter, we describe an application to analyze microscopic slide data, such as might be found in medical examinations of patient samples or forensic evidence from a crime scene. We illustrate how a Hadoop system might be used to organize, analyze, and correlate bioinformatics data.

■ **Note**   This chapter uses a freely available set of fruit fly images to show how microscope images can be analyzed. Strictly speaking, these images are coming from an electron microscope, which enables a much higher magnification and resolution of the images than the ordinary optical microscope you probably first encountered in high school biology. The principles of distributed analytics on a sensors data output is the same, however. You might, for example, use images from a small drone aircraft and perform analytics on the images output from the drone camera. The software components and many of the analytical operations remain the same.

11.1   Introduction to Bioinformatics

Biology has had a long history as a science, spanning many centuries. Yet, only in the last fifty years or so has biological data used as computer data come into its own as a way of understanding the information.

Bioinformatics is the understanding of biological data as computer data, and the disciplined analysis of that computer data. We perform bioinformatics by leveraging specialized libraries to translate and validate the information contained in biological and medical data sets, such as x-rays, images of microscope slides, chemical and DNA analysis, sensor information such as cardiograms, MRI data, and many other kinds of data sources.

The optical microscope has been around for hundreds of years, but it is only relatively recently that microscope slide images have been analyzed by image processing software. Initially, these analyses were performed in a very ad-hoc fashion. Now, however, microscope slide images have become "big data" sets in their own right, and can be analyzed by using a data analytics pipeline as we've been describing throughout the book.

In this chapter, we examine a distributed analytics system specifically designed to perform the automated microscope slide analysis we saw diagrammed in Figure 8-1. As in our other examples, we will use standard third-party libraries to build our analytical system on top of Apache Hadoop and Spark infrastructure.

© Kerry Koitzsch 2017

K. Koitzsch, *Pro Hadoop Data Analytics*, DOI 10.1007/978-1-4842-1910-2_11

For an in-depth description of techniques and algorithms for medical bioinformatics, see Kalet (2009).

Before we dive into the example, we should re-emphasize the point made in the node earlier in the introduction. Whether we use electron microscopy images, optical images of a microscope slide, or even more complex images such as the DICOM images that typically represent X-rays.

■ **Note**   Several domain-specific software components are required in this case study, and include some packages specifically designed to integrate microscopes and their cameras into a standard image-processing application.

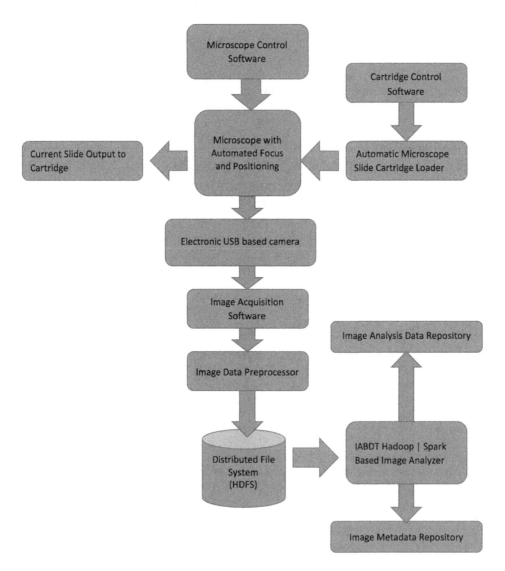

Figure 11-1. *A microscope slide analytics example with software and hardware components*

The sample code example we will discuss in this chapter is based on the architecture shown in Figure 11-1. Mostly we're not concerned with the physical mechanics of the mechanisms, unless we want fine control over the microscope's settings. The analytics system begins where the image acquisition part of the process ends. As with all of our sample applications, we go through a simple technology stack–assembling phase before we begin to work on our customized code. Working with microscopes is a special case of image processing, that is, "images as big data," which we will discuss in more detail in Chapter 14.

As we select software components for our technology stack, we also evolve the high-level diagram of what we want to accomplish in software. One result of this thinking might look like Figure 11-2. We have data sources (which essentially come from the microscope camera or cameras), processing elements, analytics elements, and result persistence. Some other components, such as a cache repository to hold intermediate results, are also necessary.

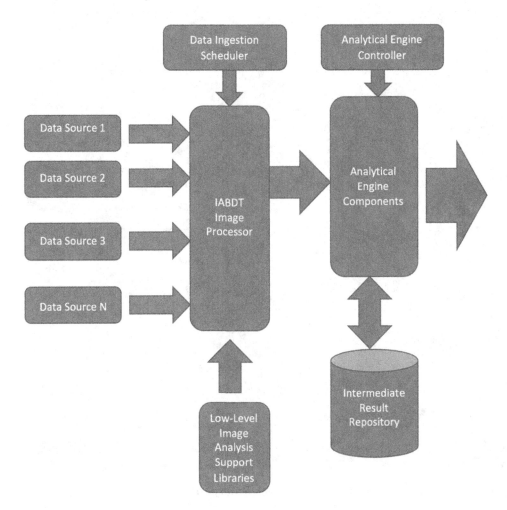

Figure 11-2. A microscope slide software architecture: high-level software component diagram

11.2   Introduction to Automated Microscopy

Figures 11-3 to 11-5 show the stages a slide goes through in automated microscopy.

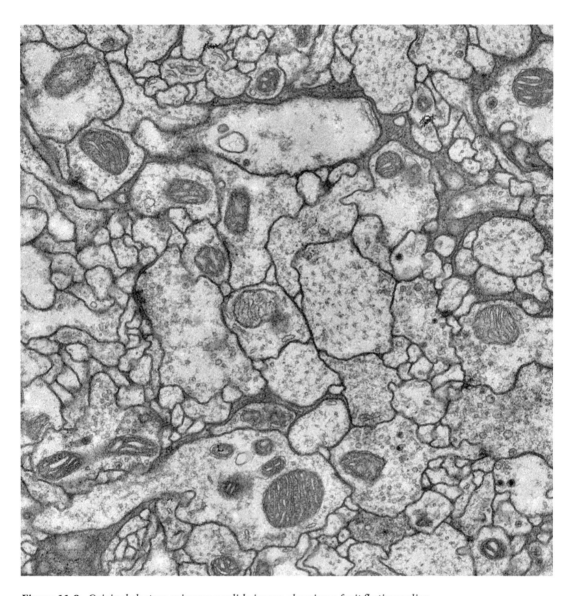

Figure 11-3. *Original electron microscope slide image, showing a fruit fly tissue slice*

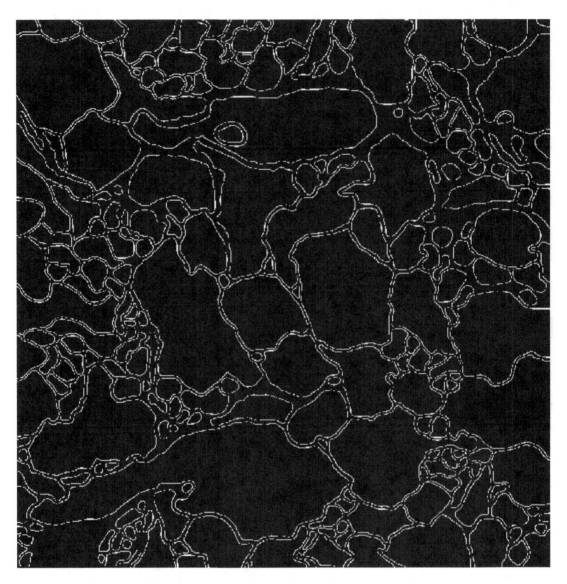

Figure 11-4. *Contour extraction from the microscope image*

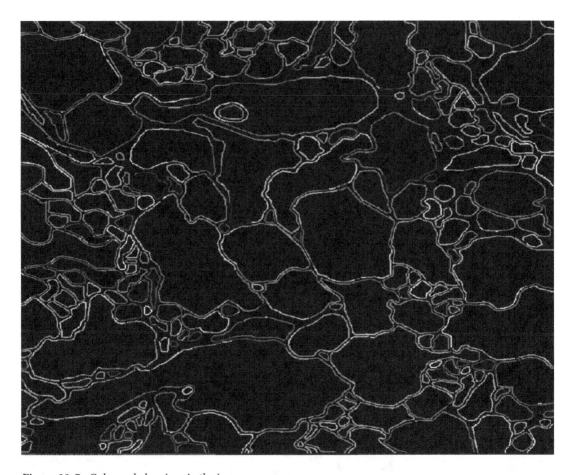

Figure 11-5. *Color-coded regions in the image*

We can use a geometric model of the tissue slices as shown in Figure 11-6.

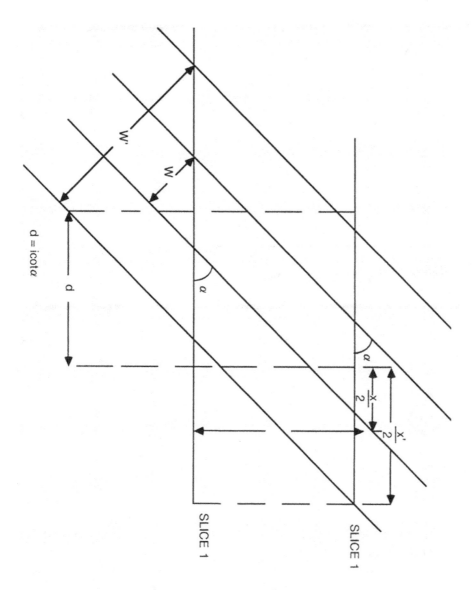

Figure 11-6. *Geometric computation of slice dimensions*

We can use three-dimensional visualization tools to analyze a stack of neural tissue slices, as shown in the examples in Figures 11-7 and 11-8.

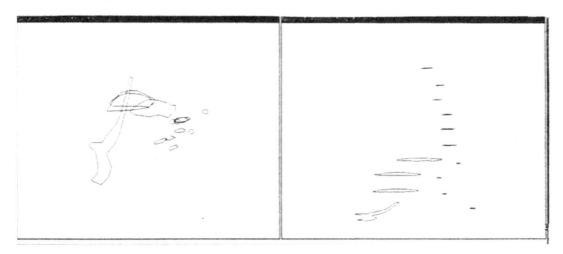

Figure 11-7. *An example of analyzing slices of neural tissue*

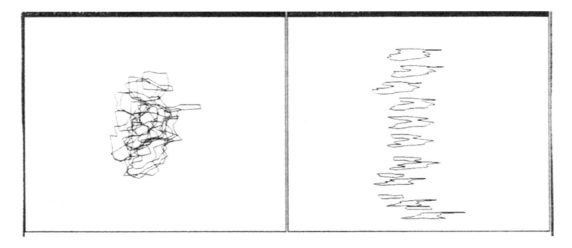

Figure 11-8. *Another example of organizing neural tissue*

11.3   A Code Example: Populating HDFS with Images

We will use the HIPI package (`http://hipi.cs.virginia.edu/gettingstarted.html`) to ingest the images into HDFS. Apache Oozie can be used to schedule the importing. We can start with a basic Hadoop job following the online instructions for HIPI:

```
package com.apress.probda.image;
import org.apache.hadoop.conf.Configured;
import org.apache.hadoop.util.Tool;
import org.apache.hadoop.util.ToolRunner;
```

```java
public class ImageProcess extends Configured implements Tool {
  public int run(String[] args) throws Exception {
    System.out.println("---- Basic HIPI Example ----");
    return 0;
  }
  public static void main(String[] args) throws Exception {
    ToolRunner.run(new ImageProcess(), args);
    System.exit(0);
  }
}
```

Edit, compile, and run the program to verify results.

The second iteration of the program is as follows:

```java
package com.apress.probda.image;
import org.hipi.image.FloatImage;
import org.hipi.image.HipiImageHeader;
import org.hipi.imagebundle.mapreduce.HibInputFormat;
import org.apache.hadoop.conf.Configured;
import org.apache.hadoop.util.Tool;
import org.apache.hadoop.util.ToolRunner;
import org.apache.hadoop.fs.Path;
import org.apache.hadoop.io.IntWritable;
import org.apache.hadoop.io.Text;
import org.apache.hadoop.mapreduce.lib.input.FileInputFormat;
import org.apache.hadoop.mapreduce.lib.output.FileOutputFormat;
import org.apache.hadoop.mapreduce.Job;
import org.apache.hadoop.mapreduce.Mapper;
import org.apache.hadoop.mapreduce.Reducer;
import org.apache.hadoop.mapreduce.lib.input.FileInputFormat;
import org.apache.hadoop.mapreduce.lib.output.FileOutputFormat;
import java.io.IOException;
public class ImageProcess extends Configured implements Tool {

  public static class ImageProcessMapper extends Mapper<HipiImageHeader, FloatImage,
IntWritable, FloatImage> {
    public void map(HipiImageHeader key, FloatImage value, Context context)
      throws IOException, InterruptedException {
    }
  }
  public static class ImageProcessReducer extends Reducer<IntWritable, FloatImage,
IntWritable, Text> {
    public void reduce(IntWritable key, Iterable<FloatImage> values, Context context)
      throws IOException, InterruptedException {
    }
  }
  public int run(String[] args) throws Exception {
    // Check input arguments
    if (args.length != 2) {
      System.out.println("Usage: imageProcess <input HIB> <output directory>");
      System.exit(0);
    }
```

```
  // Initialize and configure MapReduce job
  Job job = Job.getInstance();
  // Set input format class which parses the input HIB and spawns map tasks
  job.setInputFormatClass(HibInputFormat.class);
  // Set the driver, mapper, and reducer classes which express the computation
  job.setJarByClass(ImageProcess.class);
  job.setMapperClass(ImageProcessMapper.class);
  job.setReducerClass(ImageProcessReducer.class);
  // Set the types for the key/value pairs passed to/from map and reduce layers
  job.setMapOutputKeyClass(IntWritable.class);
  job.setMapOutputValueClass(FloatImage.class);
  job.setOutputKeyClass(IntWritable.class);
  job.setOutputValueClass(Text.class);
  // Set the input and output paths on the HDFS
  FileInputFormat.setInputPaths(job, new Path(args[0]));
  FileOutputFormat.setOutputPath(job, new Path(args[1]));
  // Execute the MapReduce job and block until it complets
  boolean success = job.waitForCompletion(true);

  // Return success or failure
  return success ? 0 : 1;
  }
  public static void main(String[] args) throws Exception {
  ToolRunner.run(new ImageProcess(), args);
  System.exit(0);
  }
}
```

Look for the complete code example in the code contributions.

```
hipi — -bash — 141×27
Kerrys-MBP:hipi kkoitzsch$ tools/hibImport.sh /Users/kkoitzsch/groundtruth-drosophila-vnc/stack1//synapses  flydata3.hib
Input image directory: /Users/kkoitzsch/groundtruth-drosophila-vnc/stack1//synapses
Input FS: local FS
Output HIB: flydata3.hib
Overwrite HIB if it exists: false
16/07/20 21:28:39 WARN util.NativeCodeLoader: Unable to load native-hadoop library for your platform... using builtin-java classes where appl
icable
 ** added: 00.png
 ** added: 01.png
 ** added: 02.png
 ** added: 03.png
 ** added: 04.png
 ** added: 05.png
 ** added: 06.png
 ** added: 07.png
 ** added: 08.png
 ** added: 09.png
 ** added: 10.png
 ** added: 11.png
 ** added: 12.png
 ** added: 13.png
 ** added: 14.png
 ** added: 15.png
 ** added: 16.png
 ** added: 17.png
 ** added: 18.png
 ** added: 19.png
```

Figure 11-9. *Successful population of HDFS with University of Virginia's HIPI system*

Check that the images have been loaded successfully with the HibInfo.sh tool by typing the following on the command line:

```
tools/hibInfo.sh flydata3.hib --show-meta
```

You should see results similar to those in Figure 11-10.

```
● ● ●                              hipi — -bash — 115×50
Kerrys-MBP:hipi kkoitzsch$ tools/hibInfo.sh flydata3.hib --show-meta
16/07/20 21:29:00 WARN util.NativeCodeLoader: Unable to load native-hadoop library for your platform... using built
in-java classes where applicable
Input HIB: flydata3.hib
Display meta data: true
Display EXIF data: false
IMAGE INDEX: 0
   1024 x 1024
   format: 2
   meta: {filename=00.png, source=/Users/kkoitzsch/groundtruth-drosophila-vnc/stack1/synapses/00.png}
IMAGE INDEX: 1
   1024 x 1024
   format: 2
   meta: {filename=01.png, source=/Users/kkoitzsch/groundtruth-drosophila-vnc/stack1/synapses/01.png}
IMAGE INDEX: 2
   1024 x 1024
   format: 2
   meta: {filename=02.png, source=/Users/kkoitzsch/groundtruth-drosophila-vnc/stack1/synapses/02.png}
IMAGE INDEX: 3
   1024 x 1024
   format: 2
   meta: {filename=03.png, source=/Users/kkoitzsch/groundtruth-drosophila-vnc/stack1/synapses/03.png}
IMAGE INDEX: 4
   1024 x 1024
   format: 2
   meta: {filename=04.png, source=/Users/kkoitzsch/groundtruth-drosophila-vnc/stack1/synapses/04.png}
IMAGE INDEX: 5
   1024 x 1024
   format: 2
   meta: {filename=05.png, source=/Users/kkoitzsch/groundtruth-drosophila-vnc/stack1/synapses/05.png}
IMAGE INDEX: 6
   1024 x 1024
   format: 2
   meta: {filename=06.png, source=/Users/kkoitzsch/groundtruth-drosophila-vnc/stack1/synapses/06.png}
IMAGE INDEX: 7
   1024 x 1024
   format: 2
   meta: {filename=07.png, source=/Users/kkoitzsch/groundtruth-drosophila-vnc/stack1/synapses/07.png}
IMAGE INDEX: 8
   1024 x 1024
   format: 2
   meta: {filename=08.png, source=/Users/kkoitzsch/groundtruth-drosophila-vnc/stack1/synapses/08.png}
IMAGE INDEX: 9
   1024 x 1024
   format: 2
   meta: {filename=09.png, source=/Users/kkoitzsch/groundtruth-drosophila-vnc/stack1/synapses/09.png}
IMAGE INDEX: 10
   1024 x 1024
   format: 2
   meta: {filename=10.png, source=/Users/kkoitzsch/groundtruth-drosophila-vnc/stack1/synapses/10.png}
```

Figure 11-10. *Successful description of HDFS images (with metadata information included)*

11.4 Summary

In this chapter, we described an example application which uses distributed bioinformatics techniques to analyze microscope slide data.

In the next chapter, we will talk about a software component based on a Bayesian approach to classification and data modeling. This turns out to be a very useful technique to supplement our distributed data analytics system, and has been used in a variety of domains including finance, forensics, and medical applications.

11.5 References

Gerhard, Stephan, Funke, Jan, Martel, Julien, Cardona, Albert, and Fetter, Richard. "Segmented anisotropic ssTEM dataset of neural tissue." Retrieved 16:09, Nov 20, 2013 (GMT) http://dx.doi.org/10.6084/m9.figshare.856713

Kalet, Ira J. *Principles of Biomedical Informatics*. London, UK: Academic Press Elsevier, 2009.

Nixon, Mark S., and Aguado, Alberto S. *Feature Extraction & Image Processing for Computer Vision, Third Edition*. London, UK: Academic Press Elsevier, 2008.

■ ■ ■

CHAPTER 12

A Bayesian Analysis Component: Identifying Credit Card Fraud

In this chapter, we describe a Bayesian analysis software component plug-in which may be used to analyze streams of credit card transactions in order to identify fraudulent use of the credit card by illicit users.

■ **Note** We will primarily use the Naïve Bayes implementation provided by Apache Mahout, but we will discuss several potential solutions to using Bayesian analysis in general.

12.1 Introduction to Bayesian Analysis

Bayesian networks (which are also known as belief networks or probabilistic causal networks) are representations of observations, experiments, or hypotheses. The whole concept of "belief" and "Bayesian network" go hand in hand. When we perform a physical experiment, such as using a Geiger counter to identify radioactive minerals, or a chemical test of a soil sample to infer the presence of natural gas, coal, or petroleum, there is a "belief factor" associated with the results of these experiments. How accurate is the experiment? How reliable is the "data model" of the experiment—its premises, data, relationships within data variables, methodology? And how much do we believe the "conclusions" of the experiment? Fortunately, a lot of the infrastructure we've built up over the last few chapters is very useful in dealing with Bayesian technologies of all kinds, especially the graph databases. Almost all Bayesian network problems benefit from being represented as graphs—after all, they are networks—and the graph database can assist with a seamless representation of Bayesian problems.

■ **Note** Bayesian analysis is a gigantic area of continually evolving concepts and technologies, which now include deep learning and machine learning aspects. Some of the references at the end of the chapter provide an overview of concepts, algorithms, and techniques, which have been used so far in Bayesian analysis.

Bayesian techniques are particularly relevant to an ongoing financial problem: the identification of credit card fraud. Let's take a look at a simple credit card fraud algorithm, as shown in Figure 18-1. The implementation and algorithm shown is based on the work of Triparthi and Ragha (2004).

We will describe how to build a distributed credit card fraud detector based on the algorithm shown in Figure 12-1, using some of the by now familiar strategies and techniques described in previous chapters.

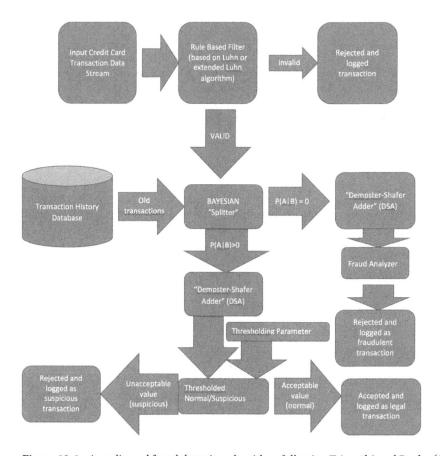

Figure 12-1. *A credit card fraud detection algorithm, following Triparthi and Ragha (2004)*

First things first: add an environment variable to your .bash_profile file for this application:

```
export CREDIT_CARD_HOME=/Users/kkoitzsch/probda/src/main/resources/creditcard
```

First, lets get some credit card test data. We start with the data sets found at https://www.cs.purdue.edu/commugrate/data/credit_card/. This data set was the basis for one of the Code Challenges of 2009. We are only interested in these files:

```
DataminingContest2009.Task2.Test.Inputs
DataminingContest2009.Task2.Train.Inputs
DataminingContest2009.Task2.Train.Targets
```

Download the files into $CREDIT_CARD_HOME/data.

Let's look at the structure of the credit card transaction record. Each line in the CSV file is a transaction record consisting of the following fields:

```
amount,hour1,state1,zip1,custAttr1,field1,custAttr2,field2,hour2,flag1,total,field3,field4,i
ndicator1,indicator2,flag2,flag3,flag4,flag5
000000000025.90,00,CA,945,1234567890197185,3,redjhmbdzmbzg1226@sbcglobal.net,0,00,0,00000000
0025.90,2525,8,0,0,1,0,0,2
000000000025.90,00,CA,940,1234567890197186,0,puwelzumjynty@aol.com,0,00,0,000000000025.90,3
393,17,0,0,1,1,0,1
000000000049.95,00,CA,910,1234567890197187,3,quhdenwubwydu@earthlink.
net,1,00,0,000000000049.95,-737,26,0,0,1,0,0,1
000000000010.36,01,CA,926,1234567890197202,2,xkjrjiokleeur@hotmail.com,0,01,1,000000000010.3
6,483,23,0,0,1,1,0,1
000000000049.95,01,CA,913,1234567890197203,3,yzlmmssadzbmj@socal.rr.c
om,0,01,0,000000000049.95,2123,23,1,0,1,1,0,1
```

…and more.

Looking at the standard structure for the CSV line in this data set, we notice something about field number 4: while it has a 16-digit credit-card-like code, it doesn't conform to a standard valid credit card number that would pass the Luhn test.

We write a program that will modify the event file to something more suitable: the fourth field of each record will now contain a "valid" Visa or Mastercard randomly generated credit card number, as shown in Figure 12-2. We want to introduce a few "bad" credit card numbers just to make sure our detector can spot them.

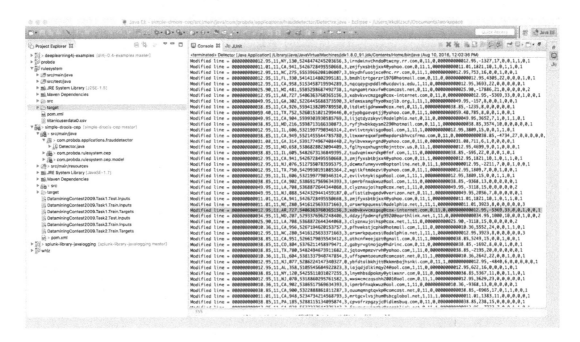

Figure 12-2. *Merging valid and invalid "real" credit card numbers with test data*

12.2   A Bayesian Component for Credit Card Fraud Detection

A Bayesian component to identify credit card fraud from data sets is the same, in principle, to many of the other kinds of data pipelines we've been discussing. It gets back to the fundamental principle of this book: distributed analytics systems are always some kind of data pipeline, some kind of workflow processing. Different arrangements, configurations, and technology choices may be used, but they share some underlying identities as far as overall design goes.

12.2.1   The Basics of Credit Card Validation

We start with the fundamental principles of credit card validation. A credit card number can be determined as valid using the Luhn check, shown in Listing 12-1.

```
public static boolean checkCreditCard(String ccNumber)
    {
            int sum = 0;
            boolean alternate = false;
            for (int i = ccNumber.length() - 1; i >= 0; i--)
            {
                    int n = Integer.parseInt(ccNumber.substring(i, i + 1));
                    if (alternate)
                    {
                            n *= 2;
                            if (n > 9)
                            {
                                    n = (n % 10) + 1;
                            }
                    }
                    sum += n;
                    alternate = !alternate;
            }
            return (sum % 10 == 0);
    }
```

The Luhn credit card number verification algorithm is shown in the flowchart in Figure 12-3.

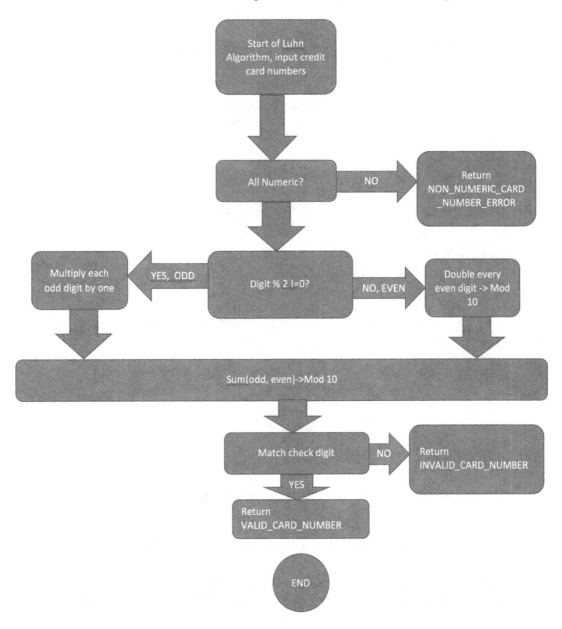

Figure 12-3. *The simple Luhn credit card validation algorithm.*

We can add machine learning techniques into the fraud-detecting mix.

Take a look at the algorithm flowchart in Figure 12-4. The process involves a training phase and a detection phase.

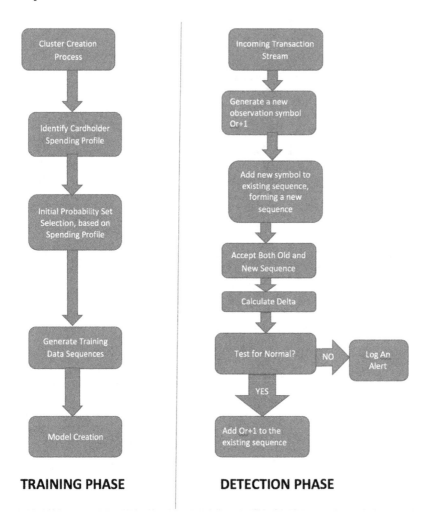

Figure 12-4. *Training and detection phases of a credit card fraud detection algorithm*

In the training phase, a clustering process creates the data model.
In the detection phase, the model previously created is used to detect (identify) new incoming events.

An implementation of the training/detecting program is shown in Figure 12-5 and Figure 12-6.

```
● ● ● 🖳 bin — java -server -Ddaemon.name=supervisor -Dstorm.options= -Dstorm.home=/Users/kkoitzsch/Downloads/apache-storm-1.0.1 -Dstorm.log.dir=/Users/kkoit...
Kerrys-MBP:bin kkoitzsch$ ./zkServer.sh start
ZooKeeper JMX enabled by default
Using config: /Users/kkoitzsch/Downloads/zookeeper-3.4.8/bin/../conf/zoo.cfg
Starting zookeeper ... STARTED
Kerrys-MBP:bin kkoitzsch$ cd ..
```

Figure 12-5. *Starting Zookeeper from the command line or script is straightforward*

```
● ● ● 🖳 bin — java -server -Ddaemon.name=supervisor -Dstorm.options= -Dstorm.home=/Users/kkoitzsch/Downloads/apache-storm-1.0.1 -Dstorm.log.dir=/Users/kkoit...
Kerrys-MBP:bin kkoitzsch$ ./storm supervisor
Running: java -server -Ddaemon.name=supervisor -Dstorm.options= -Dstorm.home=/Users/kkoitzsch/Downloads/apache-storm-1.0.1 -Dstorm.log.dir=/Users/k
koitzsch/Downloads/apache-storm-1.0.1/logs -Djava.library.path=/usr/local/lib:/opt/local/lib:/usr/lib -Dstorm.conf.file= -cp /Users/kkoitzsch/Downl
oads/apache-storm-1.0.1/lib/asm-5.0.3.jar:/Users/kkoitzsch/Downloads/apache-storm-1.0.1/lib/clojure-1.7.0.jar:/Users/kkoitzsch/Downloads/apache-sto
rm-1.0.1/lib/disruptor-3.3.2.jar:/Users/kkoitzsch/Downloads/apache-storm-1.0.1/lib/kryo-3.0.3.jar:/Users/kkoitzsch/Downloads/apache-storm-1.0.1/lib
/log4j-api-2.1.jar:/Users/kkoitzsch/Downloads/apache-storm-1.0.1/lib/log4j-core-2.1.jar:/Users/kkoitzsch/Downloads/apache-storm-1.0.1/lib/log4j-ove
r-slf4j-1.6.6.jar:/Users/kkoitzsch/Downloads/apache-storm-1.0.1/lib/log4j-slf4j-impl-2.1.jar:/Users/kkoitzsch/Downloads/apache-storm-1.0.1/lib/minl
og-1.3.0.jar:/Users/kkoitzsch/Downloads/apache-storm-1.0.1/lib/objenesis-2.1.jar:/Users/kkoitzsch/Downloads/apache-storm-1.0.1/lib/reflectasm-1.10.
1.jar:/Users/kkoitzsch/Downloads/apache-storm-1.0.1/lib/servlet-api-2.5.jar:/Users/kkoitzsch/Downloads/apache-storm-1.0.1/lib/slf4j-api-1.7.7.jar:/
Users/kkoitzsch/Downloads/apache-storm-1.0.1/lib/storm-core-1.0.1.jar:/Users/kkoitzsch/Downloads/apache-storm-1.0.1/lib/storm-rename-hack-1.0.1.jar
:/Users/kkoitzsch/Downloads/apache-storm-1.0.1/conf -Xmx256m -Dlogfile.name=supervisor.log -Dlog4jContextSelector=org.apache.logging.log4j.core.asy
nc.AsyncLoggerContextSelector -Dlog4j.configurationFile=/Users/kkoitzsch/Downloads/apache-storm-1.0.1/log4j2/cluster.xml org.apache.storm.daemon.su
pervisor
```

Figure 12-6. *Starting the Apache Storm supervisor from the command line*

You can run the complete examples from the code contribution.

12.3 Summary

In this chapter, we discussed a software component developed around a Bayesian classifier, specifically designed to identify credit card fraud in a data set. This application has been re-done and re-thought many times, and in this chapter, we wanted to showcase an implementation in which we used some of the software techniques we've already developed throughout the book to motivate our discussion.

In the next chapter, we will talk about a real-world application: looking for mineral resources with a computer simulation. "Resource finding" applications are a common type of program in which real-world data sets are mined, correlated, and analyzed to identify likely locations of a "resource," which might be anything from oil in the ground to clusters of trees in a drone image, or a particular type of cell in a microscopic slide.

12.4 References

Bolstad, William M. *Introduction to Bayesian Statistics*. New York, NY: John Wiley and Sons, Inc., 2004.

Castillo, Enrique, Gutierrez, Jose Manuel, and Hadi, Ali S. *Expert Systems and Probabilistic Network Models*. New York, NY: Springer-Verlag, 1997.

Darwiche, Adnan. *Modeling and Reasoning with Bayesian Networks*. New York, NY: Cambridge University Press, 2009.

Kuncheva, Ludmila. *Combining Pattern Classifiers: Methods and Algorithms*. Hoboken, NJ: Wiley Inter-Science, 2004.

Neapolitan, Richard E. *Probabilistic Reasoning in Expert Systems: Theory and Algorithms*. New York, NY: John Wiley and Sons, Inc., 1990.

Shank, Roger, and Riesbeck, Christopher. *Inside Computer Understanding: Five Programs Plus Miniatures*. Hillsdale, NJ: Lawrence Earlbaum Associates, 1981.

Tripathi, Krishna Kumar and Ragha, Lata. "Hybrid Approach for Credit Card Fraud Detection" in *International Journal of Soft Computing and Engineering (IJSCE)* ISSN: 2231-2307, Volume-3, Issue-4, September 2013.

■ ■ ■

CHAPTER 13

Searching for Oil: Geographical Data Analysis with Apache Mahout

In this chapter, we discuss a particularly interesting application for distributed big data analytics: using a domain model to look for likely geographic locations for valuable minerals, such as petroleum, bauxite (aluminum ore), or natural gas. We touch on a number of convenient technology packages to ingest, analyze, and visualize the resulting data, especially those well-suited for processing geolocations and other geography-related data types.

■ **Note** In this chapter we use the Elasticsearch version 2.3. This version also provides the facility to use the MapQuest map visualizations you will see throughout this chapter and elsewhere in the book.

13.1 Introduction to Domain-Based Apache Mahout Reasoning

Big data analytics have many domain-specific applications, and we can use Apache Mahout to effectively address domain-centric concerns. Sometimes the knowledge base involved in the analytical process is extremely complex; data sets may be imprecise or incomplete, or the data model might be faulty, poorly thought out, or simply inappropriate for the solution requirements. Apache Mahout, as a tried-and-true machine learning infrastructure component—and the way in which it supplies well-trusted algorithms and tools—takes some of the headache out of building domain-based systems.

A relevant example of this domain-centric application is the "resource finder" application type. This includes analytical systems which process large amounts of timestamped data (sometimes over years or decades, in fact); verifies, harmonizes, and correlates the data; and then, through the use of a domain-specific data model, computes analytics (and the resultant data visualizations which are the outputs of those analytics) to identify the location of specific "resources" (usually in the earth or in the ocean). Needless to say, timestamping, collation and curation of the data, as well as accurate processing of the geolocation data, is key towards producing accurate, relevant, and timely hypotheses, explanations, summaries, suggestions, and visualizations from such a "resource finder" system.

© Kerry Koitzsch 2017
K. Koitzsch, *Pro Hadoop Data Analytics*, DOI 10.1007/978-1-4842-1910-2_13

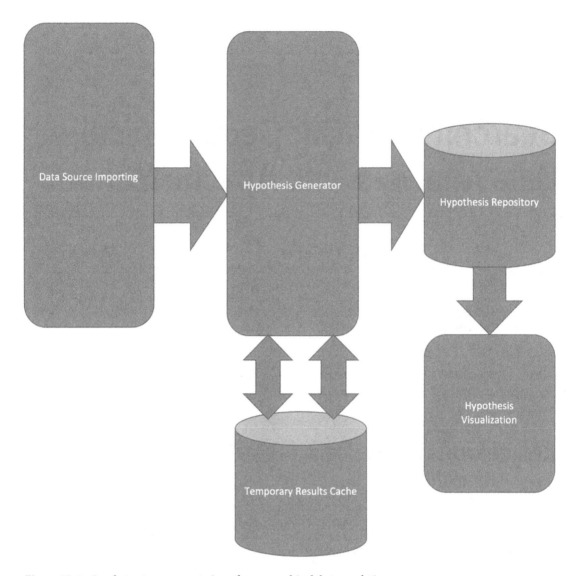

Figure 13-1. *An abstract component view of a geographical data analytics process*

In this type of system, four types of knowledge source are typically used, according to Khan "Prospector Expert System" https://www.scribd.com/doc/44131016/Prospector-Expert-System: rules (similar to those found in the JBoss Drools systems), semantic nets, and frames (a somewhat hybrid approach which is discussed thoroughly in Shank and Abelson (1981). Like other object-oriented systems, frames support inheritance, persistence, and the like.

In Figure 16.1, we show an abstracted view of a "hypothesis generator," one way in which we can predict resource locations, such as petroleum. The hypothesis generator for this example is based on JBoss Drools, which we discussed in Chapter 8.

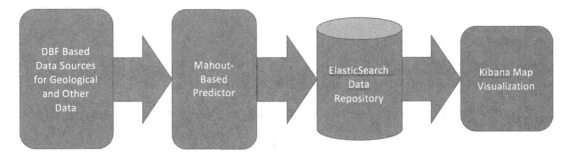

Figure 13-2. *A Mahout-based software component architecture for geographical data analysis*

In the example program, we use a DBF importer program, such as the one shown in Listing 13-1, to import data from DBF.

Elasticsearch is a very flexible data repository and a wide variety of data formats may be imported into it.

Download a few standard data sets just to get used to the Elasticsearch mechanisms. There are some samples in:

```
https://www.elastic.co/guide/en/kibana/3.0/snippets/logs.jsonl
```

as well as in

Load sample data sets just for initially testing Elasticsearch and Kibana. You can try these:

```
curl -XPOST 'localhost:9200/bank/account/_bulk?pretty' --data-binary @accounts.json
curl -XPOST 'localhost:9200/shakespeare/_bulk?pretty' --data-binary @shakespeare.json
curl -XPOST 'localhost:9200/_bulk?pretty' --data-binary @logs.jsonl
```

■ **Note**   In a previous chapter we used Apache Tika to read DBF files. In this chapter, we will use an alternative DBF reader by Sergey Polovko (Jamel). You can download this DBF reader from GitHub at `https://github.com/jamel/dbf`.

Listing 13-1. A simple DBF reader for geological data source information

```
package com.apress.probda.applications.oilfinder;

import java.io.File;
import java.util.Date;
import java.util.List;

/** We use a standard DBF reader from github.
 *
 */
import org.jamel.dbf.processor.DbfProcessor;
import org.jamel.dbf.processor.DbfRowMapper;
import org.jamel.dbf.utils.DbfUtils;
```

```java
public class Main {

        static int rownum = 0;

        public static void main(String[] args) {
        File dbf = new File("BHL_GCS_NAD27.dbf"); // pass in as args[0]

        List<OilData> oildata = DbfProcessor.loadData(dbf, new DbfRowMapper<OilData>() {
            @Override
            public OilData mapRow(Object[] row) {

                for (Object o : row) {

                        System.out.println("Row object:  " + o);

                }
                System.out.println("....Reading row: " + rownum + " into elasticsearch....");

                rownum++;

                System.out.println("------------------------");
                return new OilData(); // customize your constructor here
            }
        });

        // System.out.println("Oil Data: " + oildata);
    }
}

/** We will flesh out this information class as we develop the example.
 *
 * @author kkoitzsch
 *
 */
class OilData {

        String _name;
        int _value;
        Date _createdAt;

        public OilData(String... args){

        }

        public OilData(){

        }
```

```
public OilData(String name, int intValue, Date createdAt) {
    _name = name;
    _value = intValue;
    _createdAt = createdAt;
}

}
```

Of course, reading the geographical data (including the DBF file) is really only the first step in the analytical process.

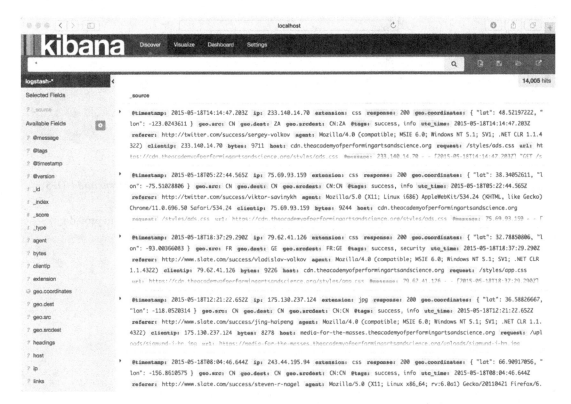

Figure 13-3. *A test query to verify Elasticsearch has been populated correctly with test data sets*

Use the Elasticsearch-Hadoop connector (https://www.elastic.co/products/hadoop) to connect Elasticsearch with Hadoop-based components of the application.

To learn more about the Hadoop-Elasticsearch connector, please refer to the web page http://www.elastic.co/guide/en/elasticsearch/hadoop/index.html.

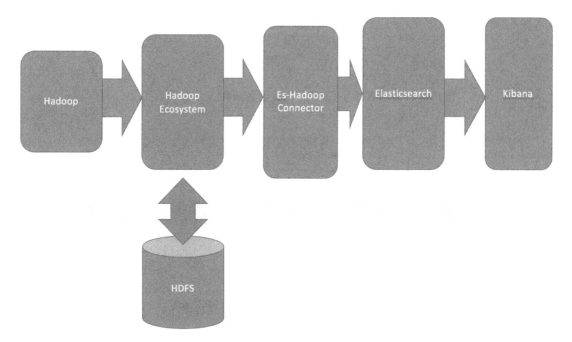

Figure 13-4. *The Elasticserch-Hadoop connector and its relationship to the Hadoop ecosystem and HDFS*

We can use the Elasticsearch-Hadoop connector in combination with SpatialHadoop to provide distributed analytic capabilities for the kind of geolocation-based data we mean to process.

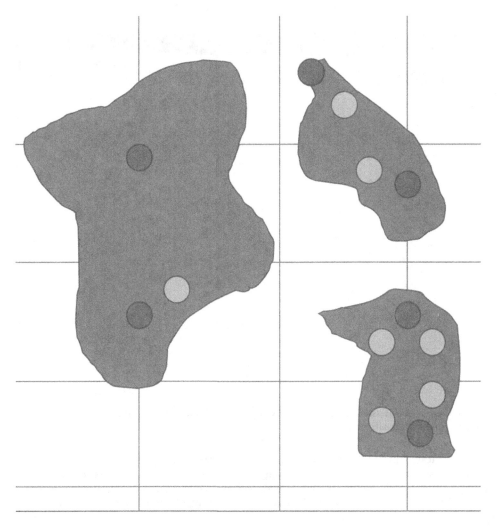

Figure 13-5. *Probability/evidence grid: a simple example of grid-based hypothesis analytic*

We can threshold our values and supply constraints on "points of interest" (spacing, how many points of interest per category, and other factors), to produce visualizations showing likelihood of desired outcomes.

Evidence and probabilities of certain desired outcomes can be stored in the same data structure, as shown in Figure 13-5. The blue regions are indicative of a likelihood that there is supporting evidence for the desired outcome, in this case, the presence of petroleum or petroleum-related products. Red and yellow circles indicate high and moderate points of interest in the hypothesis space. If the grid coordinates happen to be geolocations, one can plot the resulting hypotheses on a map similar to those shown in Figure 13-6 and Figure 13-7.

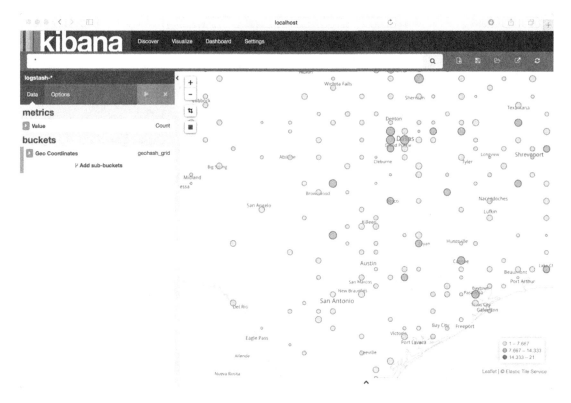

Figure 13-6. *Using Kibana and Elasticsearch for map visualiation in Texas example using latitude and logitude, and simple counts of an attribute*

We can run simple tests to insure Kibana and Elasticsearch are displaying our geolocation data correctly.

Now it is time to describe our Mahout analytical component. For this example, we will keep the analytics very simple in order to outline our thought process. Needless to say, the mathematical models of real-world resource finders would need to be much more complex, adaptable, and allow for more variables within the mathematical model.

We can use another very useful tool to prototype and view some of our data content residing in Solr using the Spatial Solr Sandbox tool by Ryan McKinley (`https://github.com/ryantxu/spatial-solr-sandbox`).

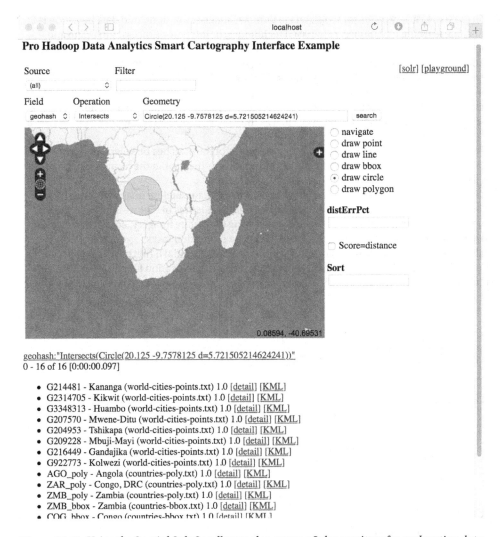

Figure 13-7. Using the Spatial Solr Sandbox tool to query a Solr repository for geolocation data

13.2 Smart Cartography Systems and Hadoop Analytics

Smart cartography (SC) systems are a special type of data pipeline–based software application which process satellite imagery, comparing the satellite images with an image database of known accuracy, called the "ground truth" database. The ground truth database provides standardized geographical location information (such as latitude and longitude of the four corners of the rectangular image, image resolution, scale, and orientation parameters) as well as other information aiding the matching process.

SC systems provide useful image match feedback to the human evaluation team, and can assist engineers and quality assurance personnel to interactively view, validate, edit, annotate, and compare incoming satellite imagery with "ground truth" imagery and metadata. Use of an SC system can enable a small team of analysts to perform the work of a much larger evaluation team in a shorter amount of time with more accurate results, because of the elimination of human error due to fatigue, observation errors, and the like.

231

SC systems can use a variety of sensor types, image formats, image resolutions, and data ingestion rates, and may use machine learning techniques, rule-based techniques, or inference processes to refine and adapt feature identification for more accurate and efficient matching between satellite image features, such as locations (latitude longitude information), image features (such as lakes, roads, airstrips, or rivers), and man-made objects (such as buildings. shopping centers, or airports).

Users of an SC system may provide feedback as to the accuracy of the computed match, which in turn allows the matching process to become more accurate over time as refinement takes place. The system may operate specifically on features selected by the user, such as the road network or man-made features such as buildings.

Finally, the SC matching process provides accuracy measures of the matches between images and ground truth data, as well as complete error and outlier information to the user in the form of reports or dashboard displays.

SC systems can provide an efficient and cost-effective way to evaluate satellite imagery for quality, accuracy, and consistency within an image sequence, and can address issues of high-resolution accuracy, task time to completion, scalability, and near real-time processing of satellite imagery, as well as providing a high-performance software solution for a variety of satellite image evaluation tasks.

One useful component to include in geolocation-centric systems is Spatial4j (`https://github.com/locationtech/spatial4j`), a helper library which provides spatial and geolocation functionality for Java programs, evolved from some of the earlier work such as the Spatial Solr Sandbox toolkit discussed earlier.

```
                                    spatial4j — -bash ▸ java — 141×48

-------------------------------------------------------
  T E S T S
-------------------------------------------------------
objc[56439]: Class JavaLaunchHelper is implemented in both /Library/Java/JavaVirtualMachines/jdk1.8.0_91.jdk/Contents/Home/jre/bin/java and /
Library/Java/JavaVirtualMachines/jdk1.8.0_91.jdk/Contents/Home/jre/lib/libinstrument.dylib. One of the two will be used. Which one is undefin
ed.
Running org.locationtech.spatial4j.context.jts.JtsSpatialContextTest
Tests run: 1, Failures: 0, Errors: 0, Skipped: 0, Time elapsed: 0.091 sec - in org.locationtech.spatial4j.context.jts.JtsSpatialContextTest
Running org.locationtech.spatial4j.context.SpatialContextFactoryTest
Tests run: 5, Failures: 0, Errors: 0, Skipped: 0, Time elapsed: 0.001 sec - in org.locationtech.spatial4j.context.SpatialContextFactoryTest
Running org.locationtech.spatial4j.distance.TestDistances
Tests run: 10, Failures: 0, Errors: 0, Skipped: 0, Time elapsed: 0.49 sec - in org.locationtech.spatial4j.distance.TestDistances
Running org.locationtech.spatial4j.io.BinaryCodecTest
Tests run: 4, Failures: 0, Errors: 0, Skipped: 0, Time elapsed: 0.008 sec - in org.locationtech.spatial4j.io.BinaryCodecTest
Running org.locationtech.spatial4j.io.GeneralGeoJSONTest
Tests run: 27, Failures: 0, Errors: 0, Skipped: 0, Time elapsed: 0.07 sec - in org.locationtech.spatial4j.io.GeneralGeoJSONTest
Running org.locationtech.spatial4j.io.GeneralPolyshapeTest
Tests run: 11, Failures: 0, Errors: 0, Skipped: 0, Time elapsed: 0.019 sec - in org.locationtech.spatial4j.io.GeneralPolyshapeTest
Running org.locationtech.spatial4j.io.GeneralWktTest
Tests run: 11, Failures: 0, Errors: 0, Skipped: 3, Time elapsed: 0.023 sec - in org.locationtech.spatial4j.io.GeneralWktTest
Running org.locationtech.spatial4j.io.JtsBinaryCodecTest
Tests run: 5, Failures: 0, Errors: 0, Skipped: 0, Time elapsed: 0.015 sec - in org.locationtech.spatial4j.io.JtsBinaryCodecTest
Running org.locationtech.spatial4j.io.JtsPolyshapeParserTest
Tests run: 1, Failures: 0, Errors: 0, Skipped: 0, Time elapsed: 0 sec - in org.locationtech.spatial4j.io.JtsPolyshapeParserTest
Running org.locationtech.spatial4j.io.JtsWKTReaderShapeParserTest
Tests run: 3, Failures: 0, Errors: 0, Skipped: 0, Time elapsed: 0.012 sec - in org.locationtech.spatial4j.io.JtsWKTReaderShapeParserTest
Running org.locationtech.spatial4j.io.JtsWktShapeParserTest
Tests run: 17, Failures: 0, Errors: 0, Skipped: 0, Time elapsed: 0.052 sec - in org.locationtech.spatial4j.io.JtsWktShapeParserTest
Running org.locationtech.spatial4j.io.LegacyShapeReadWriterTest
Tests run: 6, Failures: 0, Errors: 0, Skipped: 0, Time elapsed: 0.006 sec - in org.locationtech.spatial4j.io.LegacyShapeReadWriterTest
Running org.locationtech.spatial4j.io.ShapeFormatTest
Tests run: 2, Failures: 0, Errors: 0, Skipped: 0, Time elapsed: 0.009 sec - in org.locationtech.spatial4j.io.ShapeFormatTest
Running org.locationtech.spatial4j.io.TestGeohashUtils
Tests run: 6, Failures: 0, Errors: 0, Skipped: 0, Time elapsed: 0.002 sec - in org.locationtech.spatial4j.io.TestGeohashUtils
Running org.locationtech.spatial4j.io.WktCustomShapeParserTest
Tests run: 11, Failures: 0, Errors: 0, Skipped: 0, Time elapsed: 0.017 sec - in org.locationtech.spatial4j.io.WktCustomShapeParserTest
Running org.locationtech.spatial4j.io.WktShapeParserTest
Tests run: 9, Failures: 0, Errors: 0, Skipped: 0, Time elapsed: 0.009 sec - in org.locationtech.spatial4j.io.WktShapeParserTest
Running org.locationtech.spatial4j.shape.BufferedLineStringTest
Laps: 3800 CWIDbD: 41,106,752,75,2826
Tests run: 1, Failures: 0, Errors: 0, Skipped: 0, Time elapsed: 0.147 sec - in org.locationtech.spatial4j.shape.BufferedLineStringTest
Running org.locationtech.spatial4j.shape.BufferedLineTest
Laps: 1557 CWIDbD: 39,48,407,72,991
Tests run: 18, Failures: 0, Errors: 0, Skipped: 0, Time elapsed: 0.057 sec - in org.locationtech.spatial4j.shape.BufferedLineTest
Running org.locationtech.spatial4j.shape.impl.BBoxCalculatorTest
Tests run: 100, Failures: 0, Errors: 0, Skipped: 0, Time elapsed: 0.075 sec - in org.locationtech.spatial4j.shape.impl.BBoxCalculatorTest
Running org.locationtech.spatial4j.shape.JtsGeometryTest
```

Figure 13-8. *Running the tests for Spatial4j, a commonly used geolocation java toolkit library*

Another useful software library to use is SpatialHadoop (http://spatialhadoop.cs.umn.edu), a MapReduce-based extension to Hadoop itself. SpatialHadoop provides spatial data types, indexes, and operations which allow the use of a simple high-level language to control the processing of geolocation-centric data with Hadoop-based programs.

```
⊖ ○ ○              🗀 spatialhadoop — bash — 80×15
hadoop jar spatialhadoop-2-b2.jar generate bigdata0.dat mbr:0,0,6000,4000 20000
size:20000 —overwrite
Unable to find a $JAVA_HOME at "/usr", continuing with system-provided Java...
Unable to find a $JAVA_HOME at "/usr", continuing with system-provided Java...
14/02/18 13:20:24 WARN spatialHadoop.CommandLineArguments: unknown shape type: n
ull
Generating a file with sindex:null file of size: 20000
To: bigdata0.dat
In the range: Rectangle: (0.0,0.0)-(6000.0,4000.0)
2014-02-18 13:20:25.015 java[1947:f0f] Unable to load realm info from SCDynamicS
tore
14/02/18 13:20:28 WARN util.NativeCodeLoader: Unable to load native-hadoop libra
ry for your platform... using builtin-java classes where applicable
Generation time: 19 millis
unknown4c8d79e9253a:spatialhadoop kerryk$ ▊
```

Figure 13-9. *Generating a data file for use with SpatialHadoop*

13.3 Summary

In this chapter, we talked about the theory and practice of searching for oil and other natural resources using big data analytics as a tool. We were able to load DBF data, manipulate and analyze the data with Mahout=based code, and output the results to a simple visualizer. We also talked about some helpful libraries to include in any geolocation-centric application, such as Spatial4j and SpatialHadoop.

In the next chapter, we will talk about a particularly interesting area of big data analytics: using images and their metadata as a data source for our analytical pipeline.

13.4 References

Gheorghe, Radu, Hinman, Matthew Lee, and Russo, Roy. *Elasticsearch in Action.* Sebastopol, CA: O'Reilly Publishing, 2015.

Giacomelli, Piero. *Apache Mahout Cookbook.* Birmingham, UK: PACKT Publishing, 2013.

Sean Owen, Robin Anil, Ted Dunning, and Ellen Friedman. *Mahout in Action.* Shelter Island, NY: Manning Publications, 2011.

CHAPTER 14

■ ■ ■

"Image As Big Data" Systems: Some Case Studies

In this chapter, we will provide a brief introduction to an example toolkit, the Image as Big Data Toolkit (IABDT), a Java-based open source framework for performing a wide variety of distributed image processing and analysis tasks in a scalable, highly available, and reliable manner. IABDT is an image processing framework developed over the last several years in response to the rapid evolution of big data technologies in general, but in particular distributed image processing technologies. IABDT is designed to accept many formats of imagery, signals, sensor data, metadata, and video as data input.

A general architecture for image analytics, big data storage, and compression methods for imagery and image-derived data is discussed, as well as standard techniques for image-as-big-data analytics. A sample implementation of our image analytics architecture, IABDT addresses some of the more frequently encountered challenges experienced by the image analytics developer, including importing images into a distributed file system or cache, image preprocessing and feature extraction, applying the analysis and result visualization. Finally, we showcase some of the features of IABDT, with special emphasis on display, presentation, reporting, dashboard building, and user interaction case studies to motivate and explain our design and methodology stack choices.

14.1 An Introduction to Images as Big Data

Rapid changes in the evolution of "big data" software techniques have made it possible to perform image analytics (the automated analysis and interpretation of complex semi-structured and unstructured data sets derived from computer imagery) with much greater ease, accuracy, flexibility, and speed than has been possible before, even with the most sophisticated and high-powered single computers or data centers. The "big data processing paradigm," including Hadoop, Apache Spark, and distributed computing systems, have enabled a host of application domains to benefit from image analytics and the treatment of images as big data, including medical, aerospace, geospatial analysis, and document processing applications. Modular, efficient, and flexible toolkits are still in formative or experimental development. Integration of image processing components, data flow control, and other aspects of image analytics remain poorly defined and tentative. The rapid changes in big data technologies have made even the selection of a "technology stack" to build image analytic applications problematic. The need to solve these challenges in image analytics application development have led us to develop an architecture and baseline framework implementation specifically for distributed big data image analytics support.

In the past, low-level image analysis and machine learning modules were combined within a computational framework to accomplish domain-specific tasks. With the advent of distributed processing frameworks such as Hadoop and Apache Spark, it has been possible to build integrated image frameworks that connect seamlessly with other distributed frameworks and libraries, and in which the "image as big data" concept has become a fundamental principle of the framework architecture.

© Kerry Koitzsch 2017
K. Koitzsch, *Pro Hadoop Data Analytics*, DOI 10.1007/978-1-4842-1910-2_14

Our example toolkit IABDT provides a flexible, modular architecture which is plug-in-oriented. This makes it possible to combine many different software libraries, toolkits, systems, and data sources within one integrated, distributed computational framework. IABDT is a Java- and Scala-centric framework, as it uses both Hadoop and its ecosystem as well as the Apache Spark framework with its ecosystem to perform the image processing and image analytics functionality.

IABDT may be used with NoSQL databases such as MongoDB, Neo4j, Giraph, or Cassandra, as well as with more traditional relational database systems such as MySQL or Postgres, to store computational results and serve as data repositories for intermediate data generated by pre- and post-processing stages in the image processing pipeline. This intermediate data might consist of feature descriptors, image pyramids, boundaries, video frames, ancillary sensor data such as LIDAR, or metadata. Software libraries such as Apache Camel and Spring Framework may be used as "glue" to integrate components with one another.

One of the motivations for creating IABDT is to provide a modular extensible infrastructure for performing preprocessing, analysis, as well as visualization and reporting of analysis results—specifically for images and signals. They leverage the power of distributed processing (as with the Apache Hadoop and Apache Spark frameworks) and are inspired by such toolkits as OpenCV, BoofCV, HIPI, Lire, Caliph, Emir, Image Terrier, Apache Mahout, and many others. The features and characteristics of these image toolkits are summarized in Table 14-1. IABDT provides frameworks, modular libraries, and extensible examples to perform big data analysis on images using efficient, configurable, and distributed data pipelining techniques.

Table 14-1. *Mainstream image processing toolkit features and characteristics*

Toolkit Name	Location	Implementation Language	Description
OpenCV	opencv.org	many Language bindings, including Java	general programmatic image processing toolkit
BoofCV	boofcv.org	Java	Java-based image processing toolkit
HIPI	hipi.cs.virginia.edu	Java	image processing for Hadoop toolkit
LIRE/CALIPH/EMIR	semanticmetadata.net	Java	image searching toolkits and libraries using Lucene
ImageTerrier	imageterrier.org	Java	image indexing and search based using Lucene search engine
Java Advanced Imaging	oracle.com/ technetwork/java/ javase/overview/in...	Java	general purpose image processing toolkit, venerable but still useful

Image as Big Data toolkits and components are becoming resources in an arsenal of other distributed software packages based on Apache Hadoop and Apache Spark, as shown in Figure 14-1.

Figure 14-1. *Image as Big Data tookits as distributed systems*

Some of the distributed implementations of the module types in Figure 14-1 which are implemented in IABDT include:

Genetic Systems. There are many genetic algorithms particularly suited to image analytics[1], including techniques for sampling a large solution space, feature extraction, and classification. The first two categories of technique are more applicable to the image pre-processing and feature extraction phases of the analytical process and distributed classification techniques—even those using multiple classifiers.

Bayesian Techniques. Bayesian techniques include the naïve Bayesian algorithm found in most machine learning toolkits, but also much more.

Hadoop Ecosystem Extensions. New extensions can be built on top of existing Hadoop components to provide customized "image as big data" functionality.

Clustering, Classification, and Recommendation. These three types of analytical algorithms are present in most standard libraries, including Mahout, MLib, and H2O, and they form the basis for more complex analytical systems.

Hybrid systems integrate a lot of disparate component types into one integrated whole to perform a single function. Typically hybrid systems contain a control component, which might be a rule-based system such as Drools, or other standard control component such as Oozie, which might be used for scheduling tasks or other purposes, such as Luigi for Python (`https://github.com/spotify/luigi`)), which comes with built-in Hadoop support. If you want to try Luigi out, install Luigi using Git, and clone it into a convenient subdirectory:

```
git clone
```

```
https://github.com/spotify/luigi?cm_mc_uid=02629589701314462628476&cm_mc_
sid_50200000=1457296715
```

```
cd to the bin directory and start the server
```

```
./luigid
```

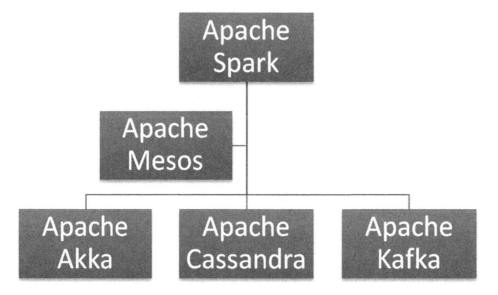

Figure 14-2. *Image as Big Data tookits as distributed systems*

14.2   First Code Example Using the HIPI System

In this section, we will introduce the HIPI Hadoop image processing system and show some simple examples of how it can be used as a distributed data processing pipeline component for images.

HIPI (hipi.cs.virginia.edu) Is a very useful Hadoop-based image processing tool, which originated at the University of Virginia. It integrates with more mainstream standard image processing libraries such as OpenCV to provide a wide palette of image processing and analytic techniques in a Hadoop-centric way.

Several basic tools for basic Hadoop-cenric image processing tasks are included with the HIPI system. These include tools to create "HIB" files (HIPI image bundles) as shown used in the diagram Figure 14-3.

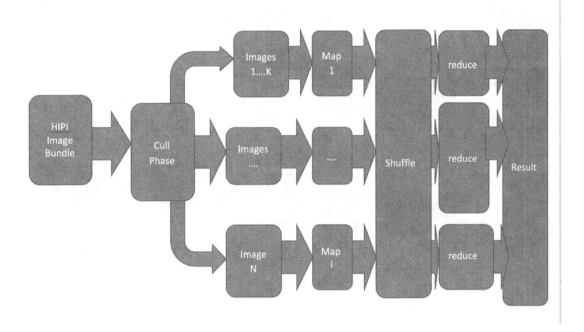

Figure 14-3. *A HIPI image data flow, consisting of bunding, culling, map/shuffle and reduce to end result*

HIPI image bundle, or "HIB," is the structured storage method used by HIPI to group images into one physical unit. The cull phase allows each HIB to be filtered out based on appropriate programmatic criteria. Images that are culled out are not fully decoded, making the HIPI pipeline much more efficient. The output of the cull phase results in image sets as shown in the diagram. Each image set has its own map phase, followed by a shuffle phase and corresponding reduce steps to create the final result. So, as you can see, the HIPI data flow is similar to the standard map-reduce data flow process. We reproduce the Hadoop data flow process in Figure 14-4 for your reference.

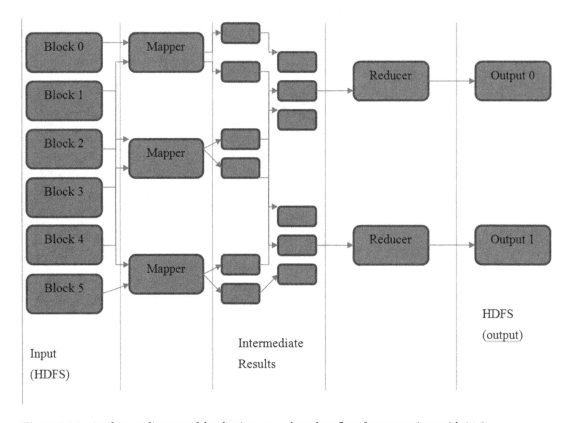

Figure 14-4. *A reference diagram of the classic map-reduce data flow, for comparison with 14-3*

INSTALLING A BASIC HIPI SYSTEM

Basic HIPI installation instructions follow.

1. First, review the "getting started" page at

    `http://hipi.cs.virginia.edu/gettingstarted.html`

    for an overview of what's in store and updates and/or changes to the system.

2. Install the basic HIPI software as shown in the "Getting Started" page:

```
git clone git@github.com:uvagfx/hipi.git
```

This will install the source code into a "hipi" directory. Cd to this "hipi" directory and "ls" the contents to review. You will need a Gradle build tool installation to install from the source. The resulting build will appear similar to Figure 14-5.

```
● ● ●                          hipi — bash — 86×46
Kerrys-MBP:~ kerryk$ cd hipi
Kerrys-MBP:hipi kerryk$ ls
README.md        core            settings.gradle util
build.gradle     license.txt     testdata         web
conf             release         tools
Kerrys-MBP:hipi kerryk$ gradle
Unable to find a $JAVA_HOME at "/usr", continuing with system-provided Java...
:core:compileJava UP-TO-DATE
:core:processResources UP-TO-DATE
:core:classes UP-TO-DATE
:core:jar UP-TO-DATE
:tools:covar:compileJava UP-TO-DATE
:tools:covar:processResources UP-TO-DATE
:tools:covar:classes UP-TO-DATE
:tools:covar:jar UP-TO-DATE
:tools:hibDownload:compileJava UP-TO-DATE
:tools:hibDownload:processResources UP-TO-DATE
:tools:hibDownload:classes UP-TO-DATE
:tools:hibDownload:jar UP-TO-DATE
:tools:hibDump:compileJava UP-TO-DATE
:tools:hibDump:processResources UP-TO-DATE
:tools:hibDump:classes UP-TO-DATE
:tools:hibDump:jar UP-TO-DATE
:tools:hibImport:compileJava UP-TO-DATE
:tools:hibImport:processResources UP-TO-DATE
:tools:hibImport:classes UP-TO-DATE
:tools:hibImport:jar UP-TO-DATE
:tools:hibInfo:compileJava UP-TO-DATE
:tools:hibInfo:processResources UP-TO-DATE
:tools:hibInfo:classes UP-TO-DATE
:tools:hibInfo:jar UP-TO-DATE
:tools:hibToJpeg:compileJava UP-TO-DATE
:tools:hibToJpeg:processResources UP-TO-DATE
:tools:hibToJpeg:classes UP-TO-DATE
:tools:hibToJpeg:jar UP-TO-DATE
:install

Finished building the HIPI library along with all tools and examples.

BUILD SUCCESSFUL

Total time: 3.935 secs

This build could be faster, please consider using the Gradle Daemon: https://docs.grad
le.org/2.7/userguide/gradle_daemon.html
Kerrys-MBP:hipi kerryk$ ▯
```

Figure 14-5. *Successful Gradle installation of the HIPI toolkit*

Gradle is another useful installation and build tool which is similar to Maven. Some systems, such as HIPI, are much easier to install using Gradle than with other techniques such as Maven.

```
○ ○ ○                          📁 tools — bash — 103×46
Munishs-MacBook-Pro:tools kerryk$ ./hibInfo.sh ./theImages0.hib
2016-03-06 18:32:43.371 java[36496:1903] Unable to load realm info from SCDynamicStore
16/03/06 18:32:45 WARN util.NativeCodeLoader: Unable to load native-hadoop library for your platform...
 using builtin-java classes where applicable
Input HIB: ./theImages0.hib
Display meta data: false
Display EXIF data: false
IMAGE INDEX: 0
    6000 x 4000
    format: 1
IMAGE INDEX: 1
    6000 x 4000
    format: 1
IMAGE INDEX: 2
    6000 x 4000
    format: 1
IMAGE INDEX: 3
    6000 x 4000
    format: 1
IMAGE INDEX: 4
    6000 x 4000
    format: 1


IMAGE INDEX: 5
    6000 x 4000
    format: 1
IMAGE INDEX: 6
    6000 x 4000
    format: 1
IMAGE INDEX: 7
    6000 x 4000
    format: 1
IMAGE INDEX: 8
    6000 x 4000
    format: 1
IMAGE INDEX: 9
    6000 x 4000
    format: 1
IMAGE INDEX: 10
    6000 x 4000
    format: 1
Found [11] images.
Munishs-MacBook-Pro:tools kerryk$
Munishs-MacBook-Pro:tools kerryk$
Munishs-MacBook-Pro:tools kerryk$ ▯
```

Figure 14-6. *Example using HIPI info utility: Mage info about a 10-image HIB in the HIPI system*

Installation of HIPI is only the first step, however! We have to integrate our HIPI processor with the analytical components to produce our results.

14.3   BDA Image Toolkits Leverage Advanced Language Features

The ability to use modern interpreted languages such as Python—along with interactive read-eval-print loops (REPLs) and functional programming—are features found with most modern programming languages, including Java 9, Scala, and interactive Python. IABDT uses these modern programming language features to make the system easier to use and the API code is much more succinct as a result.

IABDT integrates seamlessly with both Hadoop 2 and Apache Spark and uses standard distributed libraries such as Mahout, MLib, H20 and Sparkling Water to provide analytical functionality. One of the case studies we discuss also uses standard Java-centric statistical libraries with Hadoop, such as R and Weka.

14.4   What Exactly are Image Data Analytics?

Image data analytics apply the same general principles, patterns, and strategies of generic data analytics. The difference is the data source. We move away from the traditional ideas of analyzing numbers, line items, texts, documents, log entries, and other text-based data sources. Instead of a text-based data source, we are now dealing with data much less straightforward: the world of signals (which are essentially time series) and images (which can be two-dimensional images of color pixels with RGB values, or even more exotic image types with metadata, geolocations, and overlay information attached).

Specialized toolkits are needed to perform basic image data pipelining. At the top level, many pre-coded and customizable methods are provided to assist you. An assortment of these methods are shown in Table 14-2.

Table 14-2. *A selection of methods from the Image as Big Data Toolkit*

Method Name	Method Signature	Output Types	Description
EJRCL	EJRCL(Image, PropertySet)	ComputationResult	edges, junctions, regions, contours, and lines
createImagePyramid	imagePyramid(Image, PropertySet)	ImagePyramid	one image converted to an image pyramid, parametrically
projectBayesian	projectBayesian(ImageSet, BayesianModel, PropertySet)	BayesianResult	project an image set into a Bayesian hypothesis space
computeStatistics	computeStatistics(Image, PropertySet)	ComputationResult	basic statistics computed for single image, or over an image set or image pyramid
deepLearn	deepLearn(ImageSet, Learner, PropertySet)	LearnerResult	use standard distributed deep learning algorithms to process an image set or pyramid
multiClassify	multiclassify(ImageSet, ClassifierModel, PropertySet)	ClassifierResult	use multiple classifiers to classify an image set or image pyramid

Table 14-3. *Display methods for visualization provided by the IABDT. Most object types in the IABDT may be displayed using similar methods*

Method Name	Method Signature	Leverages Toolkit	Description
display	display(Image, PropertySet)	BoofCV	
display	display(ImagePyramid, PropertySet)	BoofCV	
display	display(ImageSet, PropertySet)	BoofCV	
display	display(Image, FeatureSet, PropertySet)	BoofCV	
display	display(Image, GeoLocationModel, PropertySet)	BoofCV	
display	display(Image, ResultSet, PropertySet)	BoofCV	

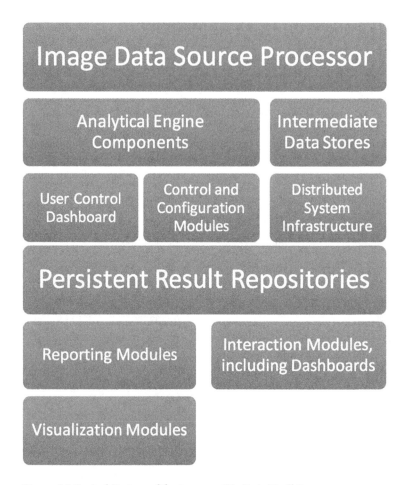

Figure 14-7. *Architecture of the Image as Big Data Toolkit*

The image data source processor is the component responsible for data acquisition, image cleansing, format conversion, and other operations to "massage" the data into formats acceptable to the other pipeline components.

The analytical engine components can be support libraries such as R and Weka.

Intermediate data sources are the outputs of initial analytical computation.

The user control dashboard is an event handler, interactive component.

The control and configuration modules consist of rule components such as Drools or other rule engine or control components, and may contain other "helper" libraries for tasks such as scheduling, data filtering and refinement, and overall system control and configuration tasks. Typically, ZooKeeper and/or Curator may be used to coordinate and orchestrate the control and configuration tasks.

The distributed system infrastructure module contains underlying support and "helper" libraries.

The persistent result repositories can be any of a variety of types of data sink, including relational, graph, or NoSQL type databases. In-memory key-value data stores may also be used if appropriate.

The reporting modules typically consist of old-school tabular or chart presentations of analytical results.

User interaction, control, and feedback is supplied by the IABDT interaction modules, which include default dashboards for common use cases.

Visualization modules consist of support libraries for displaying images, overlays, feature locations, and other visual information which make interacting and understanding the data set easier.

14.5 Interaction Modules and Dashboards

The ability to develop appropriate displays and dashboards for distributed image processing systems are an essential aid to evaluation, testing, proof-of-concept and optimization of completed implementations.

Building basic user interfaces and dashboards are supported directly in the IABDT. A picture of a simple user interface is shown in Figure 14-8.

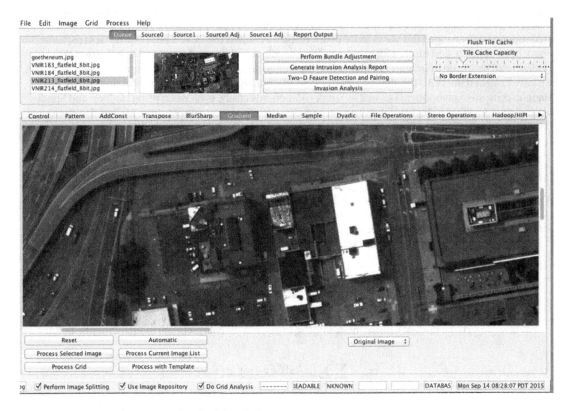

Figure 14-8. *A simple user interface build with the IABDT*

Consolidated views of the same objects, image displays which process image sequences, and image overlay capability are all provided by the IABD toolkit.

Dashboard, display, and interactive interfaces—both standalone application and web based—may be built with the IABDT user interface building module. Support for standard types of display, including overlays, and geolocation data, are provided in the prototype IABDT.

14.6 Adding New Data Pipelines and Distributed Feature Finding

Designing a new analytical dataflow using IABDT is straightforward. Equations from an algorithm may be converted into stand-alone code, and from stand-alone code to a map/reduce implementation, leveraging a number of toolkits provided for integration with the Hadoop/Spark ecosystems, including the University of Virginia's HIPI system (hipi.cs.virginia.edu), as described below.

Some distributed image processing capabilities have been explicitly developed for Spark-based systems, so a small digression on the Apache Spark vs. Hadoop controversy may be in order at this point. There has been some debate recently in the literature about whether Apache Spark has killed the map/reduce paradigm as well as "killed" the usefulness of the Hadoop ecosystem (for example, the Apache Mahout library originally started with map/reduce support only, but evolved to support Apache Spark and even H20 support). We changed our views as we evolved and developed the IABDT prototype system (Apache Spark became, more and more, a force to be reckoned with over time) and came to the realization that Hadoop and Spark are intimately complementary technologies, not at all meant to be separated. As a result, we have designed the IABDT toolkit as a modular and extremely flexible system in order that we can use Hadoop ecosystem components as well as Spark components easily, even when using Hadoop and Spark technologies together in "hybrid" dataflow development, in which components from M/R and in-memory (Hadoop and Spark) processing cooperate to provide the final results.

14.7 Example: A Distributed Feature-finding Algorithm

A distributed feature-finding algorithm may be constructed using the concept of a so-called "Hu Moment."

Hu moments are used to compute characteristic shapes.

Following Kulkani (1994), we can express the mathematics of this in the following few equations.

Standard geometric moments can be computed as follows:

$$m_{pq} = \sum_{x=-n}^{n} \sum_{y=-n}^{n} x^p y^q g(x,y)$$

Where g(x,y) is a two-dimensional index into the image g. A so-called central moment may be defined as

$$m_{pq} = \sum_{x=-n}^{n} \sum_{y=-n}^{n} (x-x')^p (y-y')^q g(x,y)$$

where x' = m_{10}/m_{00}, y' = m_{01}/m_{00}

And, when normalized for scale invariance.

$$\mu_{pq} = m_{pq}^{\gamma}$$

where

$$\gamma = \frac{(p+q)}{2} + 1$$

Rotation and scale invariant central moments can be characterized, following Hu:

$$\phi_1 = \left(\mu_{20} + \mu_{02} \right)$$

$$\phi_2 = \left(\mu_{20} - \mu_{02} \right)^2 + 4\mu_{11}^2$$

$$\phi_3 = \left(\mu_{30} - 3\mu_{12} \right)^2 + \left(3\mu_{21} - \mu_{03} \right)^2$$

$$\phi_4 = \left(\mu_{30} + \mu_{12} \right)^2 + \left(\mu_{21} + \mu_{03} \right)^2$$

$$\phi_5 = \left(\mu_{30} - 3\mu_{12} \right)\left(\mu_{30} + \mu_{12} \right)\left[\left(\mu_{30} + \mu_{12} \right)^2 - 3\left(\mu_{21} + \mu_{03} \right)^2 \right] + \\ \left(3\mu_{21} - \mu_{03} \right)\left(\mu_{21} + \mu_{03} \right)\left[3\left(\mu_{30} + \mu_{12} \right)^2 - \left(\mu_{12} + \mu_{03} \right)^2 \right]$$

$$\phi_6 = \left(\mu_{20} - \mu_{02} \right)\left[\left(\mu_{30} + \mu_{12} \right)^2 - \left(\mu_{21} + \mu_{03} \right)^2 \right] + 4\mu_{11}\left(\mu_{30} + \mu_{12} \right)\left(\mu_{21} + \mu_{03} \right)$$

$$\phi_7 = \left(3\mu_{21} - \mu_{03} \right)\left(\mu_{30} + \mu_{12} \right)\left[\left(\mu_{30} + \mu_{12} \right)^2 - 3\left(\mu_{21} + \mu_{03} \right)^2 \right] - \left(\mu_{30} - 3\mu_{12} \right)\left(\mu_{12} + \mu_{03} \right)\left[\left(3\mu_{30} + \mu_{12} \right)^2 - \left(\mu_{21} + \mu_{03} \right)^2 \right]$$

A map/reduce task in Hadoop can be coded explicitly from the moment equations, first in java for experimental purposes — to test the program logic and make sure the computed values conform to expectations — and then converted to the appropriate map/reduce constructs. A sketch of the java implementation is shown in Listing 14-1. We use a standard java class, com.apress.probda.core. ComputationalResult, to hold the answers and the "centroid" (which is also computed by our algoirithm):

Listing 14-1. Moment computation in Java

```java
public ComputationResult computeMoments(int numpoints, double[] xpts, double[] ypts)
    {
        int i;
          // array where the moments go
            double[] moments = new double[7];
        double xm.ym,x,y,xsq,ysq, factor;
        xm = ym = 0.0;
            for (i = 0; i<n; i++){
                        xm += xpts[i];
                        ym += ypts[i];
                }
        // now compute the centroid
        xm /= (double) n;
        ym /= (double) n;
        // compute the seven moments for the seven equations above
        for (i=0; i<7; i++){
        x =xpts[i]-xm;
        y = ypts[i]-ym;
        // now the seven moments
        moments[0] += (xsq=x*x); // mu 20
        moments[1] += x*y;       // mu 11
        moments[2] += (ysq = y * y); // mu 02
```

```
        moments[3] += xsq *x;        // mu 30
        moments[4] += xsq *y;        // mu 21
        moments[5] += x * ysq;       // mu 12
        moments[6] += y * ysq;       // mu 03
        }
// factor to normalize the size
        factor = 1.0 / ((double)n *(double)n);
        // second-order moment computation
        moments[0] *= factor;
        moments[1] *= factor;
        moments[2] *= factor;
        factor /= sqrt((double)n);
        // third order moment computation
        moments[3] *= factor;
        moments[4] *= factor;
        moments[5] *= factor;
        moments[6] *= factor;
        // a variety of constructors for ComputationalResult exist.
// this one constructs a result with centroid and
//moment array. ComputationResult instances are persistable.
        return new ComputationResult(xm, ym, moments);
    }
```

From this simple java implementation, we can then implement map, reduce, and combine methods with signatures such as those shown in Listing 14-2.

Listing 14-2. HIPI map/reduce method signatures for moment feature extraction computation

```
// Method signatures for the map() and reduce() methods for
// moment feature extraction module
public void map(HipiImageHeader header, FloatImage image, Context context) throws
IOException,
    InterruptedException

public void reduce(IntWritable key, Iterable momentComponents, Context context)
    throws IOException, InterruptedException
```

Lets recall the microscopy example from Chapter 11. It's a pretty typical un-structured data pipeline processing analysis problem in some ways. As you recall, image sequences start out as an ordered list of images — they may be arranged by timestamp or in more complex arrangements such as geolocation, stereo pairing, or order of importance. You can imagine in a medical application which might have dozens of medical images of the same patient, those with life-threatening anomalies should be brought to the front of the queue as soon as possible.

Other image operations might be good candidates for distributed processing, such as the Canny edge operation, coded up in BoofCV in Listing 14-3.

Listing 14-3. Canny Edge Detection Using BoofCV, before parallelization

```java
package com.apress.iabdt.examples;

import java.awt.image.BufferedImage;
import java.util.List;

import com.kildane.iabdt.model.Camera;

import boofcv.alg.feature.detect.edge.CannyEdge;
import boofcv.alg.feature.detect.edge.EdgeContour;
import boofcv.alg.filter.binary.BinaryImageOps;
import boofcv.alg.filter.binary.Contour;
import boofcv.factory.feature.detect.edge.FactoryEdgeDetectors;
import boofcv.gui.ListDisplayPanel;
import boofcv.gui.binary.VisualizeBinaryData;
import boofcv.gui.image.ShowImages;
import boofcv.io.UtilIO;
import boofcv.io.image.ConvertBufferedImage;
import boofcv.io.image.UtilImageIO;
import boofcv.struct.ConnectRule;
import boofcv.struct.image.ImageSInt16;
import boofcv.struct.image.ImageUInt8;

public class CannyEdgeDetector {

        public static void main(String args[]) {
                BufferedImage image = UtilImageIO
                                .loadImage("/Users/kerryk/Downloads/groundtruth-drosophila-
                                vnc/stack1/membranes/00.png");

                ImageUInt8 gray = ConvertBufferedImage.convertFrom(image, (ImageUInt8) null);
                ImageUInt8 edgeImage = gray.createSameShape();

                // Create a canny edge detector which will dynamically compute the
                // threshold based on maximum edge intensity
                // It has also been configured to save the trace as a graph. This is the
                // graph created while performing
                // hysteresis thresholding.

                CannyEdge<ImageUInt8, ImageSInt16> canny = FactoryEdgeDetectors.canny(2,
                true, true, ImageUInt8.class, ImageSInt16.class);
```

```
        // The edge image is actually an optional parameter. If you don't need
        // it just pass in null
        canny.process(gray, 0.1f, 0.3f, edgeImage);

        // First get the contour created by canny
        List<EdgeContour> edgeContours = canny.getContours();
        // The 'edgeContours' is a tree graph that can be difficult to process.
        // An alternative is to extract
        // the contours from the binary image, which will produce a single loop
        // for each connected cluster of pixels.
        // Note that you are only interested in external contours.
        List<Contour> contours = BinaryImageOps.contour(edgeImage,
        ConnectRule.EIGHT, null);

        // display the results
        BufferedImage visualBinary = VisualizeBinaryData.renderBinary(edgeImage,
        false, null);
        BufferedImage visualCannyContour = VisualizeBinaryData.
        renderContours(edgeContours, null, gray.width, gray.height, null);
        BufferedImage visualEdgeContour = new BufferedImage(gray.width, gray.height,
        BufferedImage.TYPE_INT_RGB);
        VisualizeBinaryData.renderExternal(contours, (int[]) null,
        visualEdgeContour);

        ListDisplayPanel panel = new ListDisplayPanel();
        panel.addImage(visualBinary, "Binary Edges");
        panel.addImage(visualCannyContour, "Canny GraphTrace");
        panel.addImage(visualEdgeContour, "Canny Binary Contours");
        ShowImages.showWindow(panel, "Image As Big Data Toolkit Canny Edge
        Extraction: ", true);
    }
}
```

Interest points are well-defined, stable image space locations which have "particular interests." For example, you might notice in Figure 14-9 that the points of interest occur at the junction points connecting other structures in the image. Corners, junctions, contours, and templates may be used to identify what we are looking for within images, and statistical analysis can be performed on the results we find.

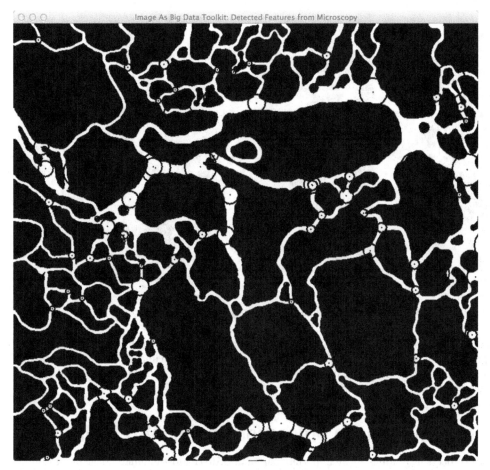

Figure 14-9. *Finding interest points in an image: the circled + signs are the interest points*

A typical input process for the IABDT is shown in Figure 14-10.

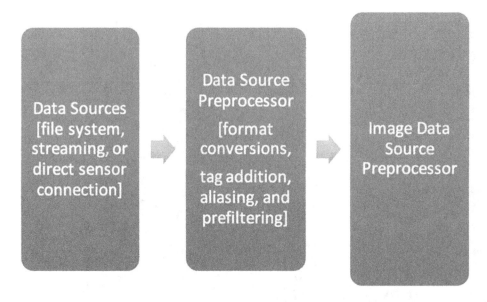

Figure 14-10. *Input process for IABD Toolkit, showing image preprocessing components*

Data sources may be processed in "batch mode" or in "streaming mode" by the data flow pipeline. The data source preprocessor is . The image data source preprocessor may perform image-centric preprocessing such as feature extraction, region identification, image pyramid construction, and other tasks to make the image processing part of the pipeline easier.

14.8   Low-Level Image Processors in the IABD Toolkit

Low-level image processing routines are an important part of the IABDT. Most standard image processing libraries, including JAI, OpenCV, and BoofCV may be used in a seamless fashion with IABDT, using Maven dependencies with the IABDT pom.xml file. Some of the standard low-level image processes included in the initial IABDT offering include Fourier operators. Fourier operators map image data into a frequency space, as shown in the following equation:

$$FP_{u,v} = \frac{1}{N}\sum_{x=0}^{N-1}\sum_{y=0}^{N-1}P_{x,y}e^{-j\left(\frac{2\pi}{N}\right)(ux+vy)}$$

Canny Edge Operators. The Canny operator can be approximated by the steps of Gaussian smoothing, the Sobel operator — a non-maximal suppression stage, thresholding (with hysteresis — a special kind of thresholding) to connecting edge points. The extracted two dimensional shapes may be persisted to an IABDT data source.

Line, Circle, and Ellipse Extraction Operators. There are feature extraction algorithms for line, circle, and ellipse shape primitives from two dimensional image data. Several sample implementations are included in the toolkit.

14.9 Terminology

Below is a brief summary of some of the terms associated with image processing and 'image as big data' concepts.

Agency-Based Systems: Cooperative multi-agent systems, or *agencies*, are an effective way to design and implement IABD systems. Individual agent node processes cooperate in a programmed network topology to achieve common goals.

Bayesian Image Processing: Array-based image processing using Bayesian techniques typically involves constructing and computing with a *Bayes network*, a graph in which the nodes are considered as random variables, and the graph edges are conditional dependencies. Random variables and conditional dependencies are standard Bayesian concepts from the fundamental Bayesian statistics. Following Opper and Winther, we can characterize Bayesian optimal prediction as

$$y^{Bayes}\left(D_m, x\right) = sgn \int df \ p\left(f \,|\, y\right) sgn \ f$$

Object hypotheses, prediction, and sensor fusion are typical problem areas for Bayesian image processing.

Classification Algorithm: Distributed classification algorithms within the IABDT include large- and small- margin (a margin is the *confidence level* of a classification) classifiers. A variety of techniques including genetic algorithms, neural nets, boosting, and support vector machines (SVMs) may be used for classification. Distributed classification algorithms, such as the standard k-means, or fuzzy-k-means techniques, are included in standard support libraries such as Apache Mahout.

Deep Learning (DL): A branch of machine learning based on learning-based data representations, and algorithms modeling high-level data abstractions. Deep learning uses multiple, complex processing levels and multiple non-linear transformations.

Distributed System: Software systems based on a messaging passing architecture over a networked hardware topology. Distributed systems may be implemented in part by software frameworks such as Apache Hadoop and Apache Spark.

Image As Big Data (IABD): The IABD concept entails treating signals, images, and video in some ways, as any other source of "big data", including the 4V conceptual basis of "variety, volume, velocity, and veracity". Special requirements for IABD include various kinds of automatic processing, such as compression, format conversion, and feature extraction.

Machine learning (ML): Machine learning techniques may be used for a variety of image processing tasks, including feature extraction, scene analysis, object detection, hypothesis generation, model building and model instantiation.

Neural net: Neural nets are a kind of mathematical model which emulate the biological models of high-level reasoning in humans. Many types of distributed neural net algorithm are useful for image analysis, feature extraction, and two- and three- dimensional model building from images.

Ontology-driven modeling: Ontologies as a description of entities within a model and the relationships between these entities, may be developed to drive and inform a modeling process, in which model refinements, metadata, and even new ontological forms and schemas, are evolved as an output of the modeling process.

Sensor fusion: Combination of information from multiple sensors or data sources into an integrated, consistent, and homogeneous data model. Sensor fusion may be accomplished by a number of mathematical techniques, including some Bayesian techniques.

Taxonomy: A scheme of classification and naming which builds a catalog. Defining, generating, or modeling a hierarchy of objects may be helped by leveraging taxonomies and related ontological data structures and processing techniques.

14.10 Summary

In this chapter, we discussed the 'image as big data' concept and why it is an important concept in the world of big data analytics techniques. The current architecture, features, and use cases for a new image-as-big-data toolkit (IABDT), was described. In it, the complementary technologies of Apache Hadoop and Apache Spark, along with their respective ecosystems and support libraries, have been unified to provide low-level image processing operations — as well as sophisticated image analysis algorithms which may be used to develop distributed, customized image processing pipelines.

In the next chapter, we discuss how to build a general-purpose data processing pipeline using many of the techniques and technology stacks we've learned from previous chapters in the book.

14.11 References

Akl, Selim G. (1989). *The Design and Analysis of Parallel Algorithms*. Englewood Cliffs, NJ: Prentice Hall.

Aloimonos, J., & Shulman, D. (1989). *Integration of Visual Modules: An Extension of the Marr Paradigm*. San Diego, CA: Academic Press Professional Inc.

Ayache, N. (1991). *Artificial Vision for Mobile Robots: Stereo Vision and Multisensory Perception*. Cambridge, MA: MIT Press.

Baggio, D., Emami, S., Escriva, D, Mahmood, M., Levgen, K., Saragih, J. (2011). *Mastering OpenCV with Practical Computer Vision Projects*. Birmingham, UK: PACKT Publications.

Barbosa, Valmir. (1996). *An Introduction of Distributed Algorithms*. Cambridge, MA: MIT Press.

Berg, M., Cheong,O., Krevald, V. M., Overmars, M. (Ed.). (2008). *Computational Geometry: Algorithms and Applications*. Berlin Heidelberg, Germany: Springer-Verlag.

Bezdek, J. C., Pal, S. K. (1992).(Ed.) *Fuzzy Models for Pattern Recognition: Methods That Search for Structures in Data*. New York, NY: IEEE Press.

Blake, A., and Yuille, A. (Ed.). (1992). *Active Vision*. Cambridge, MA: MIT Press.

Blelloch, G. E. (1990). *Vector Models for Data-Parallel Computing*. Cambridge, MA: MIT Press.

Burger, W., & Burge, M. J. (Ed.). (2016). *Digital Image Processing: An Algorithmic Introduction Using Java, Second Edition*. London, U.K. :Springer-Verlag London.

Davies, E.R. (Ed.). (2004). *Machine Vision: Theory, Algorithms, Practicalities. Third Edition*. London, U.K: Morgan Kaufmann Publishers.

Faugeras, O. (1993). *Three Dimensional Computer Vision: A Geometric Viewpoint*. Cambridge, MA: MIT Press.

Freeman, H. (Ed.) (1988*). Machine Vision: Algorithms, Architectures, and Systems*. Boston, MA: Academic Press, Inc.

Giacomelli, Piero. (2013). *Apache Mahout Cookbook*. Birmingham, UK: PACKT Publishing.

Grimson, W. E. L.; Lozano-Pez, T.;Huttenlocher, D. (1990). *Object Recognition by Computer: The Role of Geometric Constraints*. Cambridge, MA: MIT Press.

Gupta, Ashish. (2015). *Learning Apache Mahout Classification*. Birmingham, UK: PACKT Publishing.

Hare, J., Samangooei, S. and Dupplaw, D. P. (2011). *OpenIMAJ and ImageTerrier: Java libraries and tools for scalable multimedia analysis and indexing of images*. In Proceedings of the 19th ACM international conference on Multimedia (MM '11). ACM, New York, NY, USA, 691-694. DOI=10.1145/2072298.2072421 http://doi.acm.org/10.1145/2072298.2072421

Kulkarni, A. D. (1994). *Artificial Neural Networks for Image Understanding*. New York, NY: Van Nostrand Reinhold.

Kumar, V., Gopalakrishnan, P.S., Kanal, L., (Ed.). (1990). *Parallel Algorithms for Machine Intelligence and Vision*. New York NY: Springer-Verlag New York Inc.

Kuncheva, Ludmilla I. (2004). *Combining Pattern Classifiers: Methods and Algorithms*. Hoboken, New Jersey, USA: John Wiley & Sons.

Laganiere, R. (2011). *OpenCV 2 Computer Vision Application Programming Cookbook*. Birmingham, UK: PACKT Publishing.

Lindblad, T. Kinser, J.M. (Ed.). (2005*). Image Processing Using Pulse-Coupled Neural Networks, Second, Revised Edition*. Berlin Heidelberg, Germany, U.K.: Springer-Verlag Berlin Heidelberg.

Lux,M. (2015).LIRE: Lucene Image Retrieval. Retrieved on May 4, 2016 from `http://www.lire-project.net/`

Lux, M. (2015). *Caliph & Emir:MPEG-7 image annotation and retrieval GUI tools*. CaliphEmir-Caliph and Emir-Github. *In Proceedings of the 17th ACM international conference on Multimedia. ACM, 2009*. Retrieved on May 4, 2016 from `https://github.com/dermotte/CaliphEmir`

Mallat, S. (Ed.) (2009). *A Wavelet Tour of Signal Processing, The Sparse Way*. Burlington, MA, USA: Elsevier Inc.

Mallot, H. A., Allen, J.S. (Ed.). (2000). *Computational Vision: Information Processing in Perception and Visual Behavior, 2nd ed*. Cambridge, MA, USA: MIT Press.

Masters, T. (2015). *Deep Belief Nets in C++ and CUDA C. Volume 1: Restricted Boltzmann Machines*. Published by author: TimothyMasters.info

---- *Deep Belief Nets in C++ and CUDA C. Volume II: Autocoding in the Complex Domain*. (2015). Published by author: TimothyMasters.info

Nixon, M.S. and Aguado, A. S. (2012). *Feature Extraction and Image Processing for Computer Vision, Third Edition*. Oxford, U.K: Academic Press Elsevier Limited.

Pentland, Alex. (1986). *From Pixels to Predicates: Recent Advancements in Computational and Robotic Vision*. Norwood, NJ: Ablex Publishing.

Reeve, Mike (Ed.). (1989). *Parallel Processing and Artificial Intelligence*. Chichester, UK: John Wiley and Sons.

The University of Southampton (2011-2015). The ImageTerrier Image Retrieval Platform. Retrieved on May 4, 2016 from `http://www.imageterrier.org/`

Tanimoto, S, and Kilnger, A. (Eds.). (1980). *Structure Computer Vision: Machine Perception through Hierarchical Structures*. New York, NY: Academic Press.

Ullman, Shimon. (1996). *High-Level Vision: Object Recognition and Visual Cognition*. Cambridge, MA: MIT Press.

University of Virginia Computer Graphics Lab (2016). HIPI-Hadoop Image Processing Interface. Retrieved on May 4, 2016 from `http://hipi.cs.virginia.edu/`

CHAPTER 15

■ ■ ■

Building a General Purpose Data Pipeline

In this chapter, we detail an end-to-end analytical system using many of the techniques we discussed throughout the book to provide an evaluation system the user may extend and edit to create their own Hadoop data analysis system. Five basic strategies to use when developing data pipelines are discussed. Then, we see how these strategies may be applied to build a general purpose data pipeline component.

15.1   Architecture and Description of an Example System

We built some basic data pipelines in Chapter 5. Now the time has come to extend the ideas we touched on into a more general purpose data pipelining application.

Please recall that the simplest data pipeline resembles Figure 15-1. It is a series of data processing stages linked by data transmission steps. The data transmission steps collect data from a data source and emit it to a data sink. The method of transmission might be different for different transmission steps, and the data processing stages perform transformation on data inputs, emitting a data output to the subsequent stages. The final output is output to a data store or visualization/reporting component.

Figure 15-1. *A simple abstraction of a general purpose data pipeline*

Let's look at a more real-world example of a general purpose data pipeline. One of the simplest useful configurations is shown in Figure 15-2. It consists of a data source (in this case HDFS), a processing element (in this case Mahout), and an output stage (in this case a D3 visualizer which is part of the accompanying Big Data Toolkit).

© Kerry Koitzsch 2017   257
K. Koitzsch, *Pro Hadoop Data Analytics*, DOI 10.1007/978-1-4842-1910-2_15

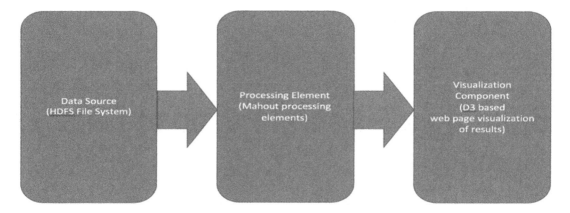

Figure 15-2. *A real-world distributed pipeline can consist of three basic elements*

Our first example imports a data set into HDFS, performs some simple analytics processing using Mahout, and passes the results of the analysis to a simple visualization component.

15.2   How to Obtain and Run the Example System

The example system is a Maven-based Java/Scala-centric system similar to many of the software components described throughout this book, and is available on the Apress code contribution site. See Appendix A and B for further details. Installation of this example system is straightforward: just follow the instructions included with the software download. Use of the infrastructure tools such as Java, Ant, and Maven have all been thoroughly described throughout the book, although the version numbers of the components may have changed. You can easily update version numbers within the pom.xml Maven file for your project.

15.3   Five Strategies for Pipeline Building

Most of this book has referred to the different strategies of data pipeline building. While software components, platforms, tools, and libraries may change, the fundamental strategic design methods of data pipeline design remain the same.

There are many strategies for data pipeline building but, broadly speaking, there are five major strategies based on "way of working." These five basic strategy types are discussed briefly below.

15.3.1   Working from Data Sources and Sinks

Working from data sources and sinks is a good organizational strategy to use when you have pre-existing or legacy data sources to use. In particular, these might include relational data, CSV flat files, or even directories full of images or log files.

When working using this data-sources-and-sinks strategy, an organized approach would include the following:

- Identify data source/sink types and provide components for data ingestion, data validation, and data cleansing (if necessary). For the purposes of this example, we will use Splunk, Tika, and Spring Framework.

- Treat the "business logic" as a black box. Initially concentrate on data input and output as well as the supporting technology stack. If the business logic is relatively simple, already packaged as a library, well-defined and straightforward to implement, we can treat the business logic component as a self-contained module or "plug-in." If the business logic requires hand-coding or is more complex

15.3.2 Middle-Out Development

Middle-out development means what it says: starting in the "middle" of the application construct and working towards either end, which in our examples will always be the data sources at the beginning of the process and the data sinks or final result repository at the end of the data pipeline. The "middle" we're developing first is essentially the "business logic" or "target algorithms" to be developed. We can start with general technology stack considerations (such as the choice to use Hadoop, Spark, or Flink, for example, or a hybrid approach using one or more of these).

15.3.3 Enterprise Integration Pattern (EIP)-based Development

EIP-based development is a useful way to develop pipelines. As we've seen, some of the standard toolkits are specifically designed to implement EIP components, and other parts of the system can be conceptualized using EIPs. Let's look at a couple of EIP diagrams to get started.

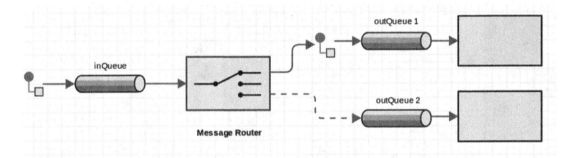

Figure 15-3. *A simple Enterprise Integration Pattern (EIP)*

We can use any of the freely available EIP diagram editors, such as the draw.io tool (draw.io) or Omnigraffle (omnigraffle.com), to draw EIP diagrams. We can then use Spring Integration or Apache Camel to implement the pipelines.

A full description of the EIP notation can be found in Hohpe and Woolf (2004).

The components shown in the abstract diagram Figure 15-4 can be implemented using Apache Camel or Spring Integration. The two endpoints are data ingestion and data persistence, respectively. The small TV screen–like symbol indicates a data visualization component and/or management console.

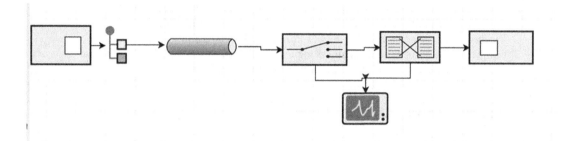

Figure 15-4. *A more extended example of an EIP*

15.3.4 Rule-based Messaging Pipeline Development

We discussed rule-based systems and how they may be used for control, scheduling, and ETL-oriented operations in Chapter 8. However, rule-based systems can be used as the center or core control mechanism of a data pipelining flow, as shown in Figure 15-5.

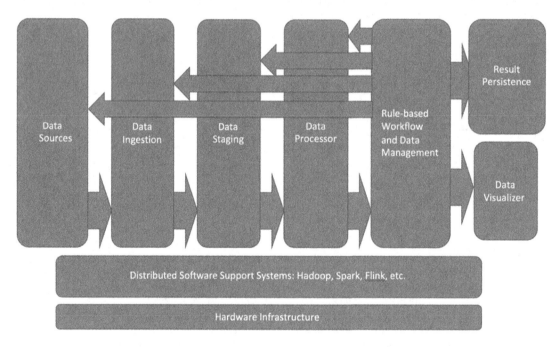

Figure 15-5. *A rule-based data flow pipeline architecture*

Figure 15-5 shows a typical architecture for a rule-based data pipeline in which all the processing components in the pipe are controlled by the rule-based workflow/data management component. Let's look at how such an architecture might be implemented.

15.3.5 Control + Data (Control Flow) Pipelining

We can essentially go back to the classic pipe-and-filter design pattern when we define a control mechanism and data stages to be controlled, as shown in the EIP diagram of Figure 15-7.

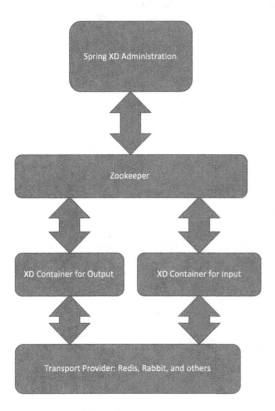

Figure 15-6. *An EIP diagram showing a different incarnation of the data pipeline*

15.4   Summary

In this chapter, we discussed construction of a general purpose data pipeline. General purpose data pipelines are an important starting point in big data analytical systems: both conceptually and in real world application building. These general purpose pipelines serve as a staging area for more application-specific extensions, as well as experimental proof-of-concept systems which may require more modification and testing before they are developed further. Starting on a strong general-purpose technology base makes it easier to perform re-work efficiently, and to "take a step back" if application requirements change.

Five basic pipeline building strategies were discussed: working from sources and sinks, middle-out development (analytical stack-centric development), enterprise integration pattern (EIP) pipeline development, rule-based messaging pipelines, and control + data (control flow) pipelining. Support libraries, techniques, and code which supports these five general purpose pipelining strategies were also discussed.

In the next and final chapter, we discuss directions for the future of big data analytics and what the future evolution of this type of system might look like.

15.5   References

Hohpe, Gregor, and Woolf, Bobby. *Enterprise Integration Patterns: Designing, Building, and Deploying Messaging Solutions.* Boston, MA: Addison-Wesley Publishing, 2004.

Ibsen, Claus, and Ansley, Jonathan. *Camel in Action.* Stamford, CT: Manning Publications, 2011.

Kavis, Michael. *Architecting the Cloud: Design Decisions for Cloud Computing Service Models.* Hoboken, NJ: John Wiley and Sons, Inc., 2014.

Mak, Gary. *Spring Recipes: A Problem-Solution Approach.* New York, NY: Springer-Verlag, Apress Publishing, 2008.

■ ■ ■

CHAPTER 16

Conclusions and the Future of Big Data Analysis

In this final chapter, we sum up what we have learned in the previous chapters and discuss some of the developing trends in big data analytics, including "incubator" projects and "young" projects for data analysis. We also speculate on what the future holds for big data analysis and the Hadoop ecosystem—"Future Hadoop" (which may also include Apache Spark and others).

Please keep in mind that the 4Vs of big data (velocity, veracity, volume, and variety) will only become larger and more complex over time. Our main conclusion: the scope, range, and effectiveness of big data analytics solutions must also continue to grow accordingly in order to keep pace with the data available!

16.1 Conclusions and a Chronology

Throughout this book we've taken a technological survey of distributed business analytical systems—specifically with Hadoop in mind—as a starting point and building block for architecture, implementation, deployment, and application. We've discussed some of the languages, toolkits, libraries, and frameworks which we have found to be the most useful ways to get new Hadoop BDAs up and running. We have tried to abide by a few strategic principles as we went along to keep things flexible and adaptable to new requirements and software components that might come along in the next few months or years.

These strategic principles include the following:

1. Use a modular design/build/test strategy to maintain software dependencies, versions, and test/integration. In our case, we use Maven and related software tools to manage builds, testing, deployment, and modular addition or subtraction of new software modules or to update versions. This doesn't mean we exclude additional necessary build tools such as Bower, Gradle, Grunt, and the like. On the contrary, all good build tools, content managers, and test frameworks should be flexible enough to work together with the others. In our experimental systems, for example, it is not uncommon to see Maven, Grunt, Bower, and Git components existing together in harmony with little friction or incompatibility.

2. Strategically select a technology stack that can be adapted for future needs and changing requirements. Keeping an architectural "vision" in mind allows system designers to work together to build and maintain a coherent technology stack, which addresses the requirements. Making good initial choices as to implementation technology is important and desirable, but having a flexible approach, in order that mistakes may be corrected, is even more desirable.

© Kerry Koitzsch 2017
K. Koitzsch, *Pro Hadoop Data Analytics*, DOI 10.1007/978-1-4842-1910-2_16

3. Be able to accommodate different programming languages appropriately, in as seamless a manner as possible. As a consequence of the need to choose a technology stack selectively, even some of the simplest applications are multi-language applications these days, and may contain Java, JavaScript, HTML, Scala, and Python components within one framework.

4. Select appropriate "glueware" for component integration, testing, and optimization/deployment. As we have seen in the examples throughout this book, "glueware" is almost as important as the components being glued! Fortunately for the developer, many components and frameworks exist for this purpose, including Spring Framework, Spring Data, Apache Camel, Apache Tika, and specialized packages such as Commons Imaging and others.

5. Last but not least, maintain a flexible and agile methodology to adapt systems to newly discovered requirements, data sets, changing technologies, and volume/complexity/quantity of data sources and sinks. Requirements will constantly change, as will support technologies. An adaptive approach saves time and rework in the long run.

In conclusion, we have come to believe that following the strategic approach to system building outlined above will assist architects, developers, and managers achieve functional business analytics systems which are flexible, scalable, and adaptive enough to accommodate changing technologies, as well as being able to process challenging data sets, build data pipelines, and provide useful and eloquent reporting capabilities, including the right data visualizations to express your results in sufficient detail.

16.2   The Current State of Big Data Analysis

In the remainder of this final chapter we will examine the current state of Hadoop and note some future possible directions and developments, speculating on "Future Hadoop"—and this, of course, includes manifestations and evolutions of distributed technology—analogous to how Apache Spark, YARN, and Hadoop 2 have been milestones in the evolution of Hadoop and its ecosystem up to the present day.

First, we have to go back to the nineteenth century.

The first rumblings of a crisis in data processing technology go back at least as far as 1880. In that year, the United States Census was calculated to take eight years to process using the techniques commonly used at that time. By 1952, the US Census was processed using the UNIVAC computer's assistance. Since then, challenges to the data processing techniques of the times have been met with one solution after another: mechanical, electronic, and semiconductor hardware solutions, paired with the evolution—and revolution—of software technologies (such as generalized programming languages), as well as organization of media (from the earliest photographs and sound recordings to the latest electronic streaming, video processing and storage, and digital media recording techniques).

In 1944, visionary and librarian Fremont Rider warned against the "information crisis"[1] (which in those days meant the number of documents physically stored in a physical library) and proposed an innovative solution which he called the "micro-card": a way of representing what we now call "metadata" on one side of a transparent microform sheet, while the individual pages of the book itself are shown on the opposite side. Rider suggested that the preservation of precious one-of-a-kind books and manuscripts from the destruction of the war that was currently raging could be achieved through the use of these "micro-cards," and now, with the "data immortality" to be found on the net in such projects as `www.archive.com`, we see the archiving of electronic books anticipated by Rider's inventions.

---

[1] In *The Scholar and the Future of the Research Library*, Fremont Rider describes his solution to the information explosion of the times. It's good reading for anyone interested in how fundamental technical problems reassert themselves in different forms over time.

We've come a long way from the perforated card and mechanical calculator, through microfilm solutions like Rider's and on to the electronic computer; but keep in mind that many computational and analytical problems remain the same. As computational power increases, data volume and availability (sensors of all kinds in great number putting out data) will require not only big data analytics, but a process of so-called "sensor fusion," in which different kinds of structured, semi-structured, and unstructured data (signals, images, and streams of all shapes and sizes) must be integrated into a common analytical picture. Drones and robot technology are two areas in which "future Hadoop" may shine, and robust sensor fusion projects are already well underway.

Statistical analysis still has its place in the world of big data analysis, no matter how advanced software and hardware components become. There will always be a place for "old school" visualization of statistics, as shown in Figure 16-1 and Figure 16-2. As for the fundamental elements of classification, clustering, feature analysis, identification of trends, commonalities, matching, etc., we can expect to see all these basic techniques recast into more and more powerful libraries. Data and metadata formats—and, most importantly, their standardization and adoption throughout the big data community—will allow us to evolve the software programming paradigms for BDAs over the next few decades.

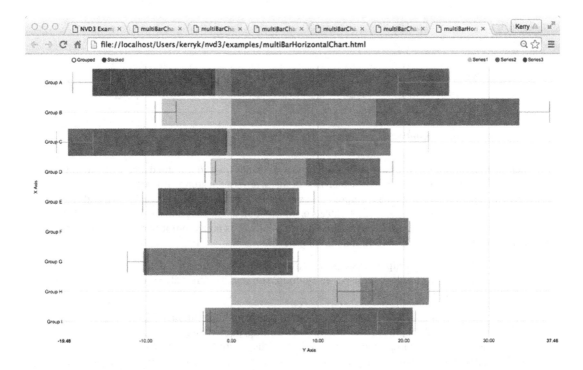

Figure 16-1. *Different kinds of "old school" bar graphs can be used to summarize grouped data*

Figure 16-2. *"Old school" candlestick graphs can still be used to summarize chronological data*

When we think about the current state of big data analysis, many questions immediately come to mind. One immediate question is, when we solve a data analytics problem, how much ground do we have to cover? What is the limit of business analytics as far as components go (keeping in mind our problem definition and scope)? Where does business analytics end and other aspects of information technology and computer science begin?

Lets take a quick review of what "business analytics" really is, as far as components go. We might start with a laundry list of components and functionalities like this:

1. Data Warehouse Components. Apache Hive started out as the go-to data warehousing technology for use with Hadoop, and is still intensively used by a vast number of software applications.

2. Business Intelligence (BI) Functionalities. The traditional definition of "Business intelligence" (BI) includes data and process mining, predictive analytics, and event processing components, but in the era of distributed BI, may also include components involving simulation, deep learning, and complex model building. BI may offer a historical, current, or predictive view of data sets, and may assist in the domain of "operational analytics," the improvement of existing operations by application of BI solutions.

3. Enterprise Integration Management (EIM). EIM is assisted by the whole area of Enterprise Integration Patterns (EIPs). Many software components, including "glueware" such as Apache Camel, are based on implementation of all or most of the EIPs found in the classic book by Hohpe and Woolf, *Enterprise Integration Patterns.*[2]

---

[2]Gregor Hohpe, Bobby Woolf. (2003) *Enterprise Integration Patterns Designing, Building, and Deploying Messaging Systems.* Addison Wesley. ISBN 978-0321200686

4. Enterprise Performance Management (EPM). EPM is an area of great interest for some vendors, particularly Cloudera. One interesting and perceptive article about this is "3 Ways 'Big Data Analytics' Will Change Enterprise Performance Management," by Bernard Marr.[3]

5. Analytic Applications (Individual Components and Functionality). Many incubating and completely new libraries and frameworks await!

6. Key Functional Requirements: Governance, Risk, and Compliance Management with Auditing.

7. Security and integrated security consistently provided throughout the core, support ecosystem, and distributed analytics application. In the early days of Hadoop development, many components within the Hadoop ecosystem had inadequate security considerations. Data provenance and monitoring-distributed systems in real time are only two of the challenges facing "future Hadoop," but they are important examples of the need for improved security measures throughout Hadoop- and Apache Spark-distributed systems.

Big data analytics capabilities will only continue to grow and prosper. Hardware and software technologies, including a new renaissance of Artificial Intelligence research and innovation, contributes to the Machine Learning and Deep Learning technologies so necessary to the further evolution of Big Data analytical techniques. Open source libraries and thriving software communities make development of new systems much more facile, even when using off-the-shelf components.

16.3 "Incubating Projects" and "Young Projects"

Throughout the book we've often referred to "mature software projects," "incubating projects," and "young projects." In this section, we'd like to take a look at what these terms mean and indicate how useful it is to architects and developers to track the "incubating" and "young" projects that might be in the queue. Please note that our examples are mostly drawn from the Apache.org web site, one of the most fertile hunting grounds for mature and maturing technology components, but there are a wide variety of other sites available for specific domain requirements. For example, a variety of image processing toolkits in various stages of maturity, development, and use is listed at `http://www.mmorph.com/resources.html` and many other similar websites.

If you take a look at the list of Apache software components on apache.org (`http://incubator.apache.org/projects/`), you'll see a host of projects either currently in incubation, graduated from incubation, and even "retired" from incubation. Graduates of the incubator go on to become full-fledged Apache projects in their own right, while retired projects may enjoy continued development even after the "retirement" event.

While the list of incubating projects is constantly changing, it's instructive to take a look at how incubating projects match up with the list of business analytics components and functionalities shown above. Some examples include Apache Atlas (`http://atlas.incubator.apache.org`) for enterprise governance services using Hadoop, using "taxonomy business annotations" for data classification. Auditing, search, lineage, and security features are provided. In contrast to the venerable Apache Hive data warehousing component, Lens (`http://incubator.apache.org/projects/lens.html`) integrates Hadoop with traditional data warehouses in a seamless manner, providing a single view of data. Lens graduated from the incubator to become a full-blown Apache project on 08-19-2015.

---

[3]`http://www.smartdatacollective.com/bernardmarr/47669/3-ways-big-data-analytics-will-change-enterprise-performance-management`
 Is the above link going to be valid as long as this book is in use?

Apache Lenya (`http://incubator.apache.org/projects/lenya.html`), a content management system, has also graduated and become an Apache project in its own right.

Security is a key concern in Hadoop-distributed systems, and several incubating components address these concerns. Some of the currently incubating projects include:

Metron (`http://incubator.apache.org/projects/metron.html`), a centralized tool for security organization and analysis, integrates a number of components from the Hadoop ecosystem to provide a scalable security analytics platform.

Ranger (`http://incubator.apache.org/projects/ranger.html`), a management framework for comprehensive data security across the Hadoop platform.

"Analytic Applications" is probably the most general category listed above in the components and capabilities list, and there are currently several incubating implementations which support component integration, dataflow construction, algorithm implementations, dashboarding, statistical analysis support, and many other necessary components of any distributed analytics application.

A few of the analytic application-centric components currently incubating at Apache include:

1. Apache Beam (`http://incubator.apache.org/projects/beam.html`) is a set of language-specific SDKs which define and execute data processing workflows as well as other types of workflows including data ingestion, integration, and others. Beam supports EIPs, (Enterprise Integration Patterns) in an analogous way to the Apache Camel system.

2. HAWQ (`http://incubator.apache.org/projects/hawq.html`) is an enterprise-quality SQL analytic engine containing an MPP (massively parallel processing) SQL framework derived from Pivotal's Greenplum Database framework. HAWQ is native to Hadoop.

3. Apache NiFi (`http://nifi.apache.org/index.html`) is a highly configurable dataflow system, a new addition to the Apache incubator as of this writing. Interestingly, NiFi provides a web-based interface to design, monitor, and control data flows.

4. MadLib (`http://madlib.incubator.apache.org`) is a big data analytic library which depends on the HAWQ SQL framework (`http://hawq.incubator.apache.org`), a "near real-time" enterprise database and query engine.

16.4   Speculations on Future Hadoop and Its Successors

Apache Hadoop has been with us for several years (2011–2016) at the time this book was written. Evolving out of the Apache Lucene and Solr search engine projects, it has taken on a life of its own and inspired potential "successors" like Apache Spark and others. What do the next steps in Hadoop core—and Hadoop ecosystem—evolution have in store for the arena of big data analytics?

One current question with Hadoop developers and architects is "Is Hadoop obsolete?" or, more precisely, "Given that Hadoop 1 has already been replaced by Hadoop 2, and Apache Spark seems to have taken the place of Hadoop in some areas, how viable is using Hadoop and its ecosystem? Are there other and perhaps better alternatives to the Hadoop ecosystem?"

We can only offer a tentative answer to these questions in this final section, basing our views on the current state of the Hadoop ecosystem and possible avenues of future development.

Please keep in mind the current functional architecture of Hadoop as shown in Figure 16-3. Let's draw a few conclusions about these based on what we learned in previous chapters.

The figure shown in 16-3 will continue to evolve and some additional components may eventually be added or subtracted over time (for example, "Programming Language Bindings" and "Backups, Disaster Recovery, and Risk Management" are obvious additions we could make, but these subjects would require book-length treatment of their own.

1. Workflow and Scheduling: Workflow and scheduling may be processed by Hadoop components like Oozie.

2. Query and Reporting Capabilities: Query and reporting capabilities could also include visualization and dashboard capabilities.

3. Security, Auditing, and Compliance: New incubating projects under the Apache umbrella address security, auditing, and compliance challenges within a Hadoop ecosystem. Examples of some of these security components include Apache Ranger (http://hortonworks.com/apache/ranger/), a Hadoop cluster security management tool.

4. Cluster Coordination: Cluster coordination is usually provided by frameworks such as ZooKeeper and library support for Apache Curator.

5. Distributed Storage: HDFS is not the only answer to distributed storage. Vendors like NetApp already use Hadoop connectors to the NFS storage system[4].

6. NoSQL Databases: As we saw in Chapter 4, there are a wide variety of NoSQL database technologies to choose from, including MongoDB and Cassandra. Graph databases such as Neo4j and Giraph are also popular NoSQL frameworks with their own libraries for data transformation, computation, and visualization.

7. Data Integration Capabilities: Data integration and glueware also continue to evolve to keep pace with different data formats, legacy programs and data, relational and NoSQL databases, and data stores such as Solr/Lucene.

8. Machine Learning: Machine learning and deep learning techniques have become an important part of the computation module of any BDAs.

9. Scripting Capabilities: Scripting capabilities in advanced languages such as Python are developing at a rapid rate, as are interactive shells or REPLs (read-eval-print loop). Even the venerable Java language includes a REPL in version 9.

10. Monitoring and System Management: The basic capabilities found in Ganglia, Nagios, and Ambari for monitoring and managing systems will continue to evolve. Some of the newer entries for system monitoring and management include Cloudera Manager (http://www.cloudera.com/products/cloudera-manager.html).

---

[4]See http://www.netapp.com/us/solutions/big-data/nfs-connector-hadoop.aspx for more information about the NetApp NFS | Hadoop Connector.

Figure 16-3. *A functional view of current Hadoop technologies and capabilities*

16.5   A Different Perspective: Current Alternatives to Hadoop

Hadoop is not the only way to go when it comes to distributed big data analytics these days. There are many alternatives to the standard Hadoop platform and ecosystem evolving today, and some of them are already supported by the Apache foundation. Please note that some of these include Apache Flink, (flink.apache. org), which is a distributed big-data analytics framework which provides an infrastructure for batch and stream data processing.

Flink can consume data from messaging systems such as Apache Kafka.

Apache Storm (storm.apache.org) is another potential competitor to Apache Spark. Storm is a real-time, distributed stream processing computation system, and supports machine learning, ETL, and "continuous computation."

16.6 Use of Machine Learning and Deep Learning Techniques in "Future Hadoop"

Algorithm implementations, "helper libraries," machine learning and deep learning libraries and frameworks, as well as improvements of classic libraries such as Mahout and MLib are important components of modern BDA implementations. They support a variety of the tasks we perform every time we construct a data processing pipeline. We should point out the differences between "machine learning" and so-called "deep learning," a relatively new term for "multiple layer neural network" algorithms, among other techniques. One very interesting example of this is the Apache Horn incubator project, described in more detail below. Machine learning, as a whole, is often seen as a "prior stage of evolution" on the road to "artificial intelligence," while "deep learning" is one step more evolved. SVMs, naïve Bayesian algorithms, and decision tree algorithms are typically called "shallow" learning techniques because, unlike deep learners, the inputs are not passed through more than one non-linear processing step before the output data set is emitted. These so-called "shallow" techniques are relegated to "machine learning," while more advanced techniques, often using many processing stages, are part of the "deep learning" repertoire.

Some of the incubating libraries at Apache (and elsewhere) which address ML and DL concerns (and their support on a variety of technology stacks and platforms) include:

1. Apache SystemML (`http://systemml.apache.org`): This library supports many standard algorithms which may be run in a distributed fashion using Hadoop or Spark. Efficient and scalable, SystemML may be run in a stand-alone mode as well as on Hadoop clusters.

2. DL4J (`http://deeplearning4j.org`): DL4J has many advanced features, such as GPU programming support.

3. H2O and Sparkling Water (`http://h2o.org`): Sparkling Water is a machine learning library based on H2O and Apache Spark. These components are programmable in Scala and feature a variety of algorithms.

4. MLib for Apache Spark (spark.apache.org): MLib is a scalable machine learning library for Spark.

5. Apache Mahout (apache.mahout.com): Mahout is a venerable and valuable part of the Hadoop ecosystem.

6. Cloudera Oryx Machine Learning Library (`https://github.com/cloudera/oryx`): Oryx is a machine learning library which supports a variety of algorithms.

7. Distributed R | Weka (`https://github.com/vertica/distributedR`, `http://weka.sourceforge.net/packageMetaData/distributedWekaHadoop/index.html`): Distributed R and Weka make a good pairing for distributed statistical analyses of all kinds, and the wide range of implemented algorithms which are available to R and Weka make implementing data pipelines much more straightforward.

8. Apache Horn (`https://horn.incubator.apache.org`): Apache Horn is an easy-to-use incubating project for deep learning. Although in its early stages of development, Apache Horn is already useful for prototyping and building useful neural net–based components for distributed analytics.

9. It's useful to see the respective features of these toolkits side-by-side, so included here is a "feature matrix" of machine learning toolkits for your reference.

271

	Naïve Bayes	Decision Trees	K Means	Deep Learning	Logistic Regression	SVM	Multi-Layer Perceptron	ALS	Linear Regression	NLP Support
Mahout	X	X	X		X		X	X	X	
H2O	X	X	X	X	X				X	
Deeplearning4j				X			X			X
MLlib		X							X	
Scikit-learn	X	X	X		X		X		X	
TensorFlow				X						
Apache Horn *										
FlinkML								X	X	
Oryx 2		X	X					X		

Figure 16-4. *A feaure matrix of ML toolkits*

* For the latest information on Apache Horn, see the incubation site at `https://horn.incubator.apache.org`.

16.7 New Frontiers of Data Visualization and BDAs

Data visualization has always been a key component of big data analytics, and with the emerging combination of graph databases and sophisticated visualization libraries such as D3, sigmajs, and many others. It has become much more straightforward to create, visualize, and interactively edit very complex—and very large—data sets. We saw some examples of this in the visualization chapter, Chapter 10. Future directions of data visualization are many, including the exploitation of holographic, virtual reality, and "telepresence" technologies. While many of these have been around for a while, distributed software systems will make sophisticated "reality systems" more and more possible as "near real-time" processing systems becomes more efficient. This will help achieve the objectives of dealing with larger and more complex data sets while maintaining compatibility, efficiency, and seamlessness with existing analytical libraries. In fact, many modern machine learning and deep learning frameworks, as well as statistical frameworks which support BDAs (such as R and Weka) contain their own visualization components and dashboards. Although some of the old-school visualization libraries were written in Java and even in C, many modern visualization libraries support a multitude of language bindings, particularly JavaScript, of course, but also Scala- and Python-based APIs, as we saw in early chapters.

16.8 Final Words

While we're considering the fate of "Future Hadoop," let's keep in mind future issues and challenges that are facing big data technologies of today.

Some of these challenges are:

1. Availability of mature predictive analytics: Being able to predict future data from existing data has always been a goal of business analytics, but much research and system building remains to be done.

2. Images and Signals as Big Data Analytics: We dived into the "images as big data" concept, in Chapter 14 and, as noted there, work is just beginning on these complex data sources, which of course include time series data and "signals" from a variety of different sensors, including LIDAR, chemical sensors for forensic analysis, medical industrial applications, accelerometer and tilt sensor data from vehicles, and many others.

3. Even Bigger Velocity, Variety and Volumes of Input Source Data: As for the required speed of data processing, the variety and level of structure and complexity (or the lack of it!), as well as raw volume of data, these requirements will become more and more demanding as hardware and software become able to deal with the increased architectural challenges and more demanding problem sets "future data analysis" will demand.

4. Combining Disparate Types of Data Sources into a Unified Analysis: "Sensor fusion" is only one aspect of combining the data into one "unified picture" of the data landscape being measured and mapped by the sensors. The evolution of distributed AI and machine learning, and the relatively new area of "deep learning," provide potential paths to moving beyond simple aggregation and fusion of different data sources by providing meaning, context, and prediction along with the raw data statistical analyses. This will enable sophisticated model building, data understanding systems, and advanced decision system applications.

5. The Merging of Artificial Intelligence (AI) and Big Data Analytics: AI, big data, and data analytics have always co-existed, even from the earliest history of AI systems. Advances in distributed machine learning (ML) and deep learning (DL) have blurred the lines between these areas even more in recent years. Deep learning libraries, such as Deeplearning4j, are routinely used in BDA applications these days, and many useful application solutions have been proposed in which AI components have been integrated seamlessly with BDAs.

6. Infrastructure and low-level support library evolution (including security): Infrastructure support toolkits for Hadoop-based applications typically include Oozie, Azkaban, Schedoscope, and Falcon. Low-level support and integration libraries include Apache Tika, Apache Camel, Spring Data and Spring Framework itself, among others. Specialized security components for Hadoop, Spark, and their ecosystems include Accumulo, Apache Sentry, Apache Knox Gateway, and many other recent contributions.

It's a good time to be in the big data analysis arena, whether you are a programmer, architect, manager, or analyst. Many interesting and game-changing future developments await. Hadoop is often seen as one stage of evolution to ever more powerful distributed analytic systems, and whether this evolution moves on to something other than "Hadoop as we know it," or the Hadoop system we already know evolves its ecosystem to process more data in better ways, distributed big data analytics is here to stay, and Hadoop is a major player in the current computing scene. We hope you have enjoyed this survey of big data analysis techniques using Hadoop as much as we have enjoyed bringing it to you.

■ ■ ■

Setting Up the Distributed Analytics Environment

This appendix is a step-by-step guide to setting up a single machine for stand-alone distributed analytics experimentation and development, using the Hadoop ecosystem and associated tools and libraries.

Of course, in a production-distributed environment, a cluster of server resources might be available to provide support for the Hadoop ecosystem. Databases, data sources and sinks, and messaging software might be spread across the hardware installation, especially those components that have a RESTful interface and may be accessed through URLs. Please see the references listed at the end of the Appendix for a thorough explanation of how to configure Hadoop, Spark, Flink, and Phoenix, and be sure to refer to the appropriate info pages online for current information about these support components.

Most of the instructions given here are hardware agnostic. The instructions are especially suited, however, for a MacOS environment.

A last note about running Hadoop based programs in a Windows environment: While this is possible and is sometimes discussed in the literature and online documentation, most components are recommended to run in a Linux or MacOS based environment.

Overall Installation Plan

The example system contains a large number of software components built around a Java-centric maven project: most of these are represented in the dependencies found in your maven pom.xml files. However, many other components are used which use other infrastructure, languages, and libraries. How you install these other components—and even whether you use them at all—is somewhat optional. Your platform may vary.

Throughout this book, as we've mentioned before, we've stuck pretty closely to a MacOS installation only. There are several reasons for this. A Mac Platform is one of the easiest environments in which to build standalone Hadoop prototypes (in the opinion of the author), and the components used throughout the book have gone through multiple versions and debugging phases and are extremely solid. Let's review the table of components that are present in the example system, as shown in Table A-1, before we discuss our overall installation plan.

© Kerry Koitzsch 2017
K. Koitzsch, *Pro Hadoop Data Analytics*, DOI 10.1007/978-1-4842-1910-2

Number	Component Name	Discussed in Chapter	URL	Description
1	Apache Hadoop	All	`hadoop.apache.org`	map/reduce distributed framework
2	Apache Spark	All	`spark.apache.org`	distributed streaming framework
3	Apache Flink	1	`flink.apache.org`	distributed stream and batch framework
4	Apache Kafka	6, 9	`kafka.apache.org`	distributed messaging framework
5	Apache Samza	9	`samza.apache.org`	distributed stream processing framework
6	Apache Gora		`gora.apache.org`	in memory data model and persistence
7	Neo4J	4	`neo4j.org`	graph database
8	Apache Giraph	4	`giraph.apache.org`	graph database
9	JBoss Drools	8	`www.drools.org`	rule framework
10	Apache Oozie		`oozie.apache.org`	scheduling component for Hadoop jobs
11	Spring Framework	All	`https://projects.spring.io/spring-framework/`	Inversion of Control Framework (IOC) and glueware
12	Spring Data	All	`http://projects.spring.io/spring-data/`	Spring Data processing (including Hadoop)
13	Spring Integration		`https://projects.spring.io/spring-integration/`	support for enterprise integration pattern-oriented programming
14	Spring XD		`http://projects.spring.io/spring-xd/`	"extreme data" integrating with other Spring components
15	Spring Batch		`http://projects.spring.io/spring-batch/`	reusable batch function library
16	Apache Cassandra		`cassandra.apache.org`	NoSQL database
17	Apache Lucene/Solr `lucene.apache.org/solr`	6	`lucene.apache.org`	open source search engine
18	Solandra	6	`https://github.com/tjake/Solandra`	Solr + Cassandra interfacing
19	OpenIMAJ	17	`openimaj.org`	image processing with Hadoop
20	Splunk	9	`splunk.com`	Java-centric logging framework
21	ImageTerrier	17	`www.imageterrier.org`	image-oriented search framework with Hadoop

(continued)

Number	Component Name	Discussed in Chapter	URL	Description
22	Apache Camel		`camel.apache.org`	general purpose glue-ware in Java: implements EIP supports
23	Deeplearning4j	12	`deeplearning4j.org`	deep learning toolkit for Java Hadoop and Spark
24 `boofcv.org`	OpenCV | BoofCV used for low-level image processing operations		`opencv.org`	
25	Apache Phoenix		`phoenix.apache.org`	OLTP and operational analytics for Hadoop
26	Apache Beam		`beam.incubator.apache.org`	unified model for creating data pipelines
27	NGDATA Lily	6	`https://github.com/NGDATA/lilyproject`	Solr and Hadoop
28	Apache Katta	6	`http://katta.sourceforge.net`	distributed Lucene with Hadoop
29	Apache Geode		`http://geode.apache.org`	distributed in-memory database
30	Apache Mahout	12	`mahout.apache.org`	machine learning library with support for Hadoop and Spark
31	BlinkDB		`http://blinkdb.org`	massively parallel, approximate query engine for running interactive SQL queries on large volumes of data.
32	OpenTSDB		`http://opentsdb.net`	time series–oriented database: runs on Hadoop and HBase
33	University of Virginia HIPI	17	`http://hipi.cs.virginia.edu/gettingstarted.html`	image processing interface with Hadoop framework
34	Distributed R and Weka statistical analysis support libraries		`https://github.com/vertica/DistributedR`	
35	Java Advanced Imaging (JAI)	17	`http://www.oracle.com/technetwork/java/download-1-0-2-140451.html`	low-level image processing package
36	Apache Kudu		`kudu.apache.org`	fast analytics processing library for the Hadoop ecosystem
37	Apache Tika		`tika.apache.org`	content-analysis toolkit

(continued)

Number	Component Name	Discussed in Chapter	URL	Description
38	Apache Apex		apex.apache.org	unified stream and batch processing framework
39	Apache Malhar		https://github.com/ apache/apex-malhar	operator and codec library for use with Apache Apex
40	MySQL Relational Database	4		
41				
42	Maven, Brew, Gradle, Gulp	All	mxaven.apache.org	build, compile, and version control infrastructure components

Once the initial basic components, such as Java, Maven, and your favorite IDE are installed, the other components may be gradually added to the system as you configure and test it, as discussed in the following sections.

Set Up the Infrastructure Components

If you develop code actively you may have some or all of these components already set up in your development environment, particularly Java, Eclipse (or your favorite IDE such as NetBeans, IntelliJ, or other), the Ant and Maven build utilities, and some other infrastructure components. The basic infrastructure components we use in the example system are listed below for your reference.

Basic Example System Setup

Set up a basic development environment. We assume that you're starting with an empty machine. You will need Java, Eclipse IDE, and Maven. These provide programming language support, an interactive development environment (IDE), and a software build and configuration tool, respectively.

First, download the appropriate Java version for development from the Oracle web site

```
http://www.oracle.com/technetwork/java/javase/downloads/jdk8-downloads-2133151.html
```

The current version of Java to use would be Java 8. Use Java-version to validate the Java version is correct. You should see something similar to Figure A-1.

```
●  ●  ●                    🏠 kkoitzsch — -bash — 80×24
Last login: Tue Aug  2 20:33:17 on ttys001
[Kerrys-MBP:~ kkoitzsch$ java -version                                    ]
java version "1.8.0_91"
Java(TM) SE Runtime Environment (build 1.8.0_91-b14)
Java HotSpot(TM) 64-Bit Server VM (build 25.91-b14, mixed mode)
Kerrys-MBP:~ kkoitzsch$ ▯
```

Figure A-1. *First step: validate Java is in place and has the correct version*

Next, download the Eclipse IDE from the Eclipse web site. Please note, we used the "Mars" version of the IDE for the development described in this book.

`http://www.eclipse.org/downloads/packages/eclipse-ide-java-ee-developers/marsr`

Finally, download the Maven-compressed version from the Maven web site `https://maven.apache.org/download.cgi`.

Validate correct Maven installation with

`mvn --version`

On the command line, you should see a result similar to the terminal output in Figure A-2.

```
●  ●  ●                    🏠 kkoitzsch — -bash — 80×24
[Kerrys-MBP:~ kkoitzsch$ mvn --version                                          ]
Apache Maven 3.3.9 (bb52d8502b132ec0a5a3f4c09453c07478323dc5; 2015-11-10T08:41:4
7-08:00)
Maven home: /Users/kkoitzsch/Downloads/apache-maven-3.3.9
Java version: 1.8.0_102, vendor: Oracle Corporation
Java home: /Library/Java/JavaVirtualMachines/jdk1.8.0_102.jdk/Contents/Home/jre
Default locale: en_US, platform encoding: UTF-8
OS name: "mac os x", version: "10.11.3", arch: "x86_64", family: "mac"
Kerrys-MBP:~ kkoitzsch$ ▯
```

Figure A-2. *Successful Maven version check*

Make sure you can log in without a passkey:

```
ssh localhost
```

If not, execute the following commands:

```
ssh-keygen -t rsa
cat ~/.ssh/id_rsa.pub >> ~/.ssh/authorized_keys
chmod 0600 ~/.ssh/authorized_keys
```

There are many online documents with complete instructions on using ssh with Hadoop appropriately, as well as several of the standard Hadoop references.

Apache Hadoop Setup

Apache Hadoop, along with Apache Spark and Apache Flink, are key components to the example system infrastructure. In this appendix we will discuss a simple installation process for each of these components. The most painless way to install these components is to refer to many of the excellent "how to" books and online tutorials about how to set up these basic components, such as Venner (2009).

Appropriate Maven dependencies for Hadoop must be added to your pom.xml file.

To configure the Hadoop system, there are several parameter files to alter as well.

Add the appropriate properties to core-site.xml:

```
<configuration>
   <property>
      <name>fs.default.name</name>
      <value>hdfs://localhost:9000</value>
   </property>
</configuration>
```

also to hdfs-site.xml:

```
<configuration>
   <property>
      <name>dfs.replication</name >
      <value>1</value>
   </property>

   <property>
      <name>dfs.name.dir</name>
      <value>file:///home/hadoop/hadoopinfra/hdfs/namenode</value>
   </property>

   <property>
      <name>dfs.data.dir</name>
      <value>file:///home/hadoop/hadoopinfra/hdfs/datanode</value>
   </property>
</configuration>
```

Install Apache Zookeeper

Download a recent release from the Zookeeper download page. Upzip the file, and add the following environment variables to the .bash_profile or equivalent file in the usual way. Please note that an installation of Zookeeper is necessary to use some of the other components, such as OpenTSDB. Review the installation instructions at https://zookeeper.apache.org/doc/trunk/zookeeperStarted.html#sc_InstallingSingleMode.

Make sure the appropriate Zookeeper environment variables are set. These include, for example:

```
export ZOOKEEPER_HOME = /Users/kkoitzsch/Downloads/zookeeper-3.4.8
```

A sample configuration file for Zookeeper is provided with the download. Place the appropriate configuration values in the file conf/zoo.cfg .

Start the Zookeeper server with the command

```
bin/zkServer.sh start
```

Check that the Zookeeper server is running with

```
ps -al | grep zook
```

You should see a response similar to the one in Figure A-3.

```
● ● ●                          ▓ bin — -bash — 80×23
[Kerrys-MBP:bin kkoitzsch$ ./zkServer.sh start                               ]
ZooKeeper JMX enabled by default
Using config: /Users/kkoitzsch/Downloads/zookeeper-3.4.8/bin/../conf/zoo.cfg
Starting zookeeper ... STARTED
[Kerrys-MBP:bin kkoitzsch$ ps -al | grep zook                               ]
  501 16464      1     4006   0  31  0  8262336  44788 -        S
  0 ttys010    0:00.47 /Library/Java/JavaVirtualMachines/jdk1.8.0_102.jdk/Content
s/Home/bin/java -Dzookeeper.log.dir=. -Dzookeeper.root.logger=INFO,CONSOLE -cp /
Users/kkoitzsch/Downloads/zookeeper-3.4.8/bin/../build/classes:/Users/kkoitzsch/
Downloads/zookeeper-3.4.8/bin/../build/lib/*.jar:/Users/kkoitzsch/Downloads/zook
eeper-3.4.8/bin/../lib/slf4j-log4j12-1.6.1.jar:/Users/kkoitzsch/Downloads/zookee
per-3.4.8/bin/../lib/slf4j-api-1.6.1.jar:/Users/kkoitzsch/Downloads/zookeeper-3.
4.8/bin/../lib/netty-3.7.0.Final.jar:/Users/kkoitzsch/Downloads/zookeeper-3.4.8/
bin/../lib/log4j-1.2.16.jar:/Users/kkoitzsch/Downloads/zookeeper-3.4.8/bin/../li
b/jline-0.9.94.jar:/Users/kkoitzsch/Downloads/zookeeper-3.4.8/bin/../zookeeper-3
.4.8.jar:/Users/kkoitzsch/Downloads/zookeeper-3.4.8/bin/../src/java/lib/*.jar:/U
sers/kkoitzsch/Downloads/zookeeper-3.4.8/bin/../conf: -Dcom.sun.management.jmxre
mote -Dcom.sun.management.jmxremote.local.only=false org.apache.zookeeper.server
.quorum.QuorumPeerMain /Users/kkoitzsch/Downloads/zookeeper-3.4.8/bin/../conf/zo
o.cfg
  501 16468 16446     4006   0  31  0  2434840    776 -        S+
  0 ttys010    0:00.00 grep zook
Kerrys-MBP:bin kkoitzsch$ []
```

Figure A-3. *Successful Zookeeper server run and process check*

Run the Zookeeper CLI (REPL) to make sure you can do simple operations using Zookeeper, as in Figure A-3.

```
● ● ●                  ▓ bin — java - zkCli.sh -server 127.0.0.1:2181 — 100×56
Kerrys-MBP:bin kkoitzsch$ ./zkCli.sh -server 127.0.0.1:2181
[Connecting to 127.0.0.1:2181
2016-10-02 11:45:02,046 [myid:] - INFO  [main:Environment@100] - Client environment:zookeeper.versio
[n=3.4.8--1, built on 02/06/2016 03:18 GMT
2016-10-02 11:45:02,049 [myid:] - INFO  [main:Environment@100] - Client environment:host.name=kerrys
-mbp.attlocal.net
2016-10-02 11:45:02,049 [myid:] - INFO  [main:Environment@100] - Client environment:java.version=1.8
.0_102
2016-10-02 11:45:02,050 [myid:] - INFO  [main:Environment@100] - Client environment:java.vendor=Orac
le Corporation
2016-10-02 11:45:02,051 [myid:] - INFO  [main:Environment@100] - Client environment:java.home=/Libra
ry/Java/JavaVirtualMachines/jdk1.8.0_102.jdk/Contents/Home/jre
2016-10-02 11:45:02,051 [myid:] - INFO  [main:Environment@100] - Client environment:java.class.path=
/Users/kkoitzsch/Downloads/zookeeper-3.4.8/bin/../build/classes:/Users/kkoitzsch/Downloads/zookeeper
-3.4.8/bin/../build/lib/*.jar:/Users/kkoitzsch/Downloads/zookeeper-3.4.8/bin/../lib/slf4j-log4j12-1.
6.1.jar:/Users/kkoitzsch/Downloads/zookeeper-3.4.8/bin/../lib/slf4j-api-1.6.1.jar:/Users/kkoitzsch/D
ownloads/zookeeper-3.4.8/bin/../lib/netty-3.7.0.Final.jar:/Users/kkoitzsch/Downloads/zookeeper-3.4.8
/bin/../lib/log4j-1.2.16.jar:/Users/kkoitzsch/Downloads/zookeeper-3.4.8/bin/../lib/jline-0.9.94.jar:
/Users/kkoitzsch/Downloads/zookeeper-3.4.8/bin/../zookeeper-3.4.8.jar:/Users/kkoitzsch/Downloads/zoo
keeper-3.4.8/bin/../src/java/lib/*.jar:/Users/kkoitzsch/Downloads/zookeeper-3.4.8/bin/../conf:
2016-10-02 11:45:02,051 [myid:] - INFO  [main:Environment@100] - Client environment:java.library.pat
h=/Users/kkoitzsch/Library/Java/Extensions:/Library/Java/Extensions:/Network/Library/Java/Extensions
:/System/Library/Java/Extensions:/usr/lib/java:.
2016-10-02 11:45:02,051 [myid:] - INFO  [main:Environment@100] - Client environment:java.io.tmpdir=/
var/folders/hz/nhnfch5j7vzdtw6vrxcrfkph0000gn/T/
2016-10-02 11:45:02,051 [myid:] - INFO  [main:Environment@100] - Client environment:java.compiler=<N
A>
2016-10-02 11:45:02,051 [myid:] - INFO  [main:Environment@100] - Client environment:os.name=Mac OS X
2016-10-02 11:45:02,051 [myid:] - INFO  [main:Environment@100] - Client environment:os.arch=x86_64
2016-10-02 11:45:02,051 [myid:] - INFO  [main:Environment@100] - Client environment:os.version=10.11
.3
2016-10-02 11:45:02,051 [myid:] - INFO  [main:Environment@100] - Client environment:user.name=kkoitz
sch
2016-10-02 11:45:02,051 [myid:] - INFO  [main:Environment@100] - Client environment:user.home=/Users
/kkoitzsch
2016-10-02 11:45:02,051 [myid:] - INFO  [main:Environment@100] - Client environment:user.dir=/Users/
kkoitzsch/Downloads/zookeeper-3.4.8/bin
2016-10-02 11:45:02,052 [myid:] - INFO  [main:ZooKeeper@438] - Initiating client connection, connect
String=127.0.0.1:2181 sessionTimeout=30000 watcher=org.apache.zookeeper.ZooKeeperMain$MyWatcher@446c
df90
Welcome to ZooKeeper!
2016-10-02 11:45:02,072 [myid:] - INFO  [main-SendThread(127.0.0.1:2181):ClientCnxn$SendThread@1032]
- Opening socket connection to server 127.0.0.1/127.0.0.1:2181. Will not attempt to authenticate us
ing SASL (unknown error)
Jline support is enabled
2016-10-02 11:45:02,137 [myid:] - INFO  [main-SendThread(127.0.0.1:2181):ClientCnxn$SendThread@876]
- Socket connection established to 127.0.0.1/127.0.0.1:2181, initiating session
[zk: 127.0.0.1:2181(CONNECTING) 0] 2016-10-02 11:45:02,202 [myid:] - INFO  [main-SendThread(127.0.0.
1:2181):ClientCnxn$SendThread@1299] - Session establishment complete on server 127.0.0.1/127.0.0.1:2
181, sessionid = 0x15786b0ab0c0000, negotiated timeout = 30000

WATCHER::

WatchedEvent state:SyncConnected type:None path:null

[zk: 127.0.0.1:2181(CONNECTED) 0] ls /
```

Figure A-4. *Zookeeper status check*

Try some simple commands in the Zookeeper CLI to insure it's functioning properly. Executing

```
ls /
create /zk_test my_data
get /zk_test
```

should show results as in Figure A-5.

```
● ● ●                        bin — java ‹ zkCli.sh -server 127.0.0.1:2181 — 100×47

[zk: 127.0.0.1:2181(CONNECTED) 0] ls /
[zookeeper]
[zk: 127.0.0.1:2181(CONNECTED) 1] create /zk_test my_data
Created /zk_test
[zk: 127.0.0.1:2181(CONNECTED) 2] ls /
[zookeeper, zk_test]
[zk: 127.0.0.1:2181(CONNECTED) 3] get /zk_test
my_data
cZxid = 0x2
ctime = Sun Oct 02 11:46:15 PDT 2016
mZxid = 0x2
mtime = Sun Oct 02 11:46:15 PDT 2016                                              ]
pZxid = 0x2
cversion = 0                                                                      ]
dataVersion = 0
aclVersion = 0                                                                    ]
ephemeralOwner = 0x0
dataLength = 7                                                                    ]
numChildren = 0
```

Figure A-5. *Successful Zookeeper process check on the command line*

Refer to `https://zookeeper.apache.org/doc/trunk/zookeeperStarted.html` for additional setup and configuration information.

Installing Basic Spring Framework Components

As with many of the Java-centric components we use, the Spring Framework components (and their ecosystem components, such as Spring XD, Spring Batch, Spring Data, and Spring Integration) essentially have two parts to their installation: downloading the sources and systems themselves, and adding the correct dependencies to the pom.xml of your Maven file. Most of the components are easy to install and use, and a consistent set of API standards runs through all the Spring components. There are also a great deal of online support, books, and Spring communities available to assist you with problem solving.

Basic Apache HBase Setup

Download a stable version of HBase from the download site in the usual way. Unzip and add the following environment variables to your .bash_profile or equivalent file.

Apache Hive Setup

Apache Hive has a few additional steps to its setup. You should have installed the basic components as shown above, including Hadoop. Download Hive from the download site, uncompress in the usual way, and set up a schema by running the schematool.

```
●●●                              🎤 kkoitzsch — -bash — 154×16
Kerrys-MBP:~ kkoitzsch$ schematool --initSchema -dbType derby                                                ]
SLF4J: Class path contains multiple SLF4J bindings.
SLF4J: Found binding in [jar:file:/Users/kkoitzsch/Downloads/apache-hive-2.1.0-bin/lib/log4j-slf4j-impl-2.4.1.jar!/org/slf4j/impl/StaticLoggerBinder.class
]
SLF4J: Found binding in [jar:file:/Users/kkoitzsch/Downloads/hadoop-2.7.2/share/hadoop/common/lib/slf4j-log4j12-1.7.10.jar!/org/slf4j/impl/StaticLoggerBin
der.class]
SLF4J: See http://www.slf4j.org/codes.html#multiple_bindings for an explanation.
SLF4J: Actual binding is of type [org.apache.logging.slf4j.Log4jLoggerFactory]
Metastore connection URL:        jdbc:derby:;databaseName=metastore_db;create=true
Metastore Connection Driver :    org.apache.derby.jdbc.EmbeddedDriver
Metastore connection User:       APP
Starting metastore schema initialization to 2.1.0
Initialization script hive-schema-2.1.0.derby.sql
Initialization script completed
schemaTool completed
Kerrys-MBP:~ kkoitzsch$ ▯
```

Figure A-6. *Successful initialization of the Hive schema using schematool*

Additional Hive Troubleshooting Tips

Some additional tips on Hive installation and troubleshooting.

Before you run Hive for the first time, run `schematool -initSchema -dbType derby`

If you already ran Hive and then tried to initSchema and it's failing:

`mv metastore_db metastore_db.tmp`

Re-run `schematool -initSchema -dbType derby`

Run Hive again.

Installing Apache Falcon

Apache Falcon architecture and use is discussed in Chapter ___.

Installing Apache Falcon can be done using the following git command on the command line:

```
git clone https://git-wip-us.apache.org/repos/asf/falcon.git falcon
cd falcon
export MAVEN_OPTS="-Xmx1024m -XX:MaxPermSize=256m -noverify" && mvn clean install
```

Installing Visualizer Software Components

This section discusses the installation and troubleshooting of user-interface and visualization software components, particularly the ones discussed in Chapter __.

Installing Gnuplot Support Software

Gnuplot is a necessary support component for OpenTSDB.

On the Mac platform, install Gnuplot using brew:

`brew install gnuplot`

on the command line. The successful result will be similar to that shown in Figure A-7.

```
● ● ●                              🔲 kkoitzsch — -bash — 97×39
==> Pouring lua-5.2.4_3.el_capitan.bottle.tar.gz
==> Caveats
Please be aware due to the way Luarocks is designed any binaries installed
via Luarocks-5.2 AND 5.1 will overwrite each other in /usr/local/bin.

This is, for now, unavoidable. If this is troublesome for you, you can build
rocks with the `--tree=` command to a special, non-conflicting location and
then add that to your `$PATH`.
==> Summary
🍺  /usr/local/Cellar/lua/5.2.4_3: 143 files, 697.3K
==> Installing gnuplot dependency: readline
==> Downloading https://homebrew.bintray.com/bottles/readline-6.3.8.el_capitan.bottle.tar.gz
######################################################################## 100.0%
==> Pouring readline-6.3.8.el_capitan.bottle.tar.gz
==> Caveats
This formula is keg-only, which means it was not symlinked into /usr/local.

OS X provides the BSD libedit library, which shadows libreadline.
In order to prevent conflicts when programs look for libreadline we are
defaulting this GNU Readline installation to keg-only.

Generally there are no consequences of this for you. If you build your
own software and it requires this formula, you'll need to add to your
build variables:

    LDFLAGS:  -L/usr/local/opt/readline/lib
    CPPFLAGS: -I/usr/local/opt/readline/include

==> Summary
🍺  /usr/local/Cellar/readline/6.3.8: 46 files, 2.0M
==> Installing gnuplot
==> Downloading https://homebrew.bintray.com/bottles/gnuplot-5.0.3.el_capitan.bottle.tar.gz
######################################################################## 100.0%
==> Pouring gnuplot-5.0.3.el_capitan.bottle.tar.gz
🍺  /usr/local/Cellar/gnuplot/5.0.3: 44 files, 2.3M
Kerrys-MBP:~ kkoitzsch$ which gnuplot
/usr/local/bin/gnuplot
Kerrys-MBP:~ kkoitzsch$ ▯
```

Figure A-7. *Successful installation of Gnuplot*

Installing Apache Kafka Messaging System

Many of the details of installing and testing the Kafka messaging system have been discussed extensively in Chapter 3. We will just touch on a few reminders here.

1. Download the Apache Kafka tar file from http://kafka.apache.org/downloads.html

2. Set the KAFKA_HOME environment variable.

3. Unzip file and go to KAFKA_HOME (in this case KAFKA_HOME would be /Users/kerryk/Downloads/kafka_2.9.1-0.8.2.2).

4. Next, start the ZooKeeper server by typing

5. bin/zookeeper-server-start.sh config/zookeeper.properties

6. Once the ZooKeeper service is up and running, type:

7. bin/kafka-server-start.sh config/server.properties

8. To test topic creation, type:

9. bin/kafka-topics.sh –create –zookeeper localhost:2181 –replication-factor 1 –partitions 1 -topic ProHadoopBDA0

10. To provide a listing of all available topics, type:

11. bin/kafka-topics.sh –list –zookeeper localhost:2181

12. At this stage, the result will be ProHadoopBDA0, the name of the topic you defined in step 5.

13. Send some messages from the console to test the messaging sending functionality. Type:

14. bin/kafka-console-producer.sh –broker-list localhost:9092 –topic ProHadoopBDA0 Now type some messages into the console.

15. You can configure a multi-broker cluster by modifying the appropriate config files. Check the Apache Kafka documentation for step-by-step processes how to do this.

Installing TensorFlow for Distributed Systems

As mentioned in the TensorFlow installation directions at `https://www.tensorflow.org/versions/r0.12/get_started/index.html`, insure that TensorFlow runs correctly by verifying that the following environment variables are set correctly:

`JAVA_HOME`: the location of your Java installation

`HADOOP_HDFS_HOME`: the location of your HDFS installation. You can also set this environment variable by running:

`source $HADOOP_HOME/libexec/hadoop-config.sh`

`LD_LIBRARY_PATH`: to include the path to `libjvm.so`. On Linux:

`Export LD_LIBRARY_PATH=$LD_LIBRARY_PATH:$JAVA_HOME/jre/lib/amd64/server`

`CLASSPATH`: The Hadoop jars must be added prior to running your TensorFlow program. The `CLASSPATH` set by `$HADOOP_HOME/libexec/hadoop-config.sh` is insufficient. Globs must be expanded as described in the `libhdfs` documentation:

`CLASSPATH=$($HADOOP_HDFS_HOME/bin/hdfs classpath --glob) python your_script.py`

Installing JBoss Drools

JBoss Drools (`http://www.drools.org`) is the core infrastructure component for rule-based scheduling and system orchestration, as well as for BPA and other purposes that we described in Chapter 8. To install JBoss Drools, download the appropriate components from the JBoss Drools download site and be sure to add the appropriate Maven dependencies to you pom.xml file. For the example system, these dependencies are already added for you.

```
● ● ●      drools-distribution-6.4.0.Final — java ‹ sh examples/runExamples.sh — 80×24
For example: ./runExamples.sh
Some notes:
- Working dir should be the directory of this script.
- Java is recommended to be JDK and java 6 for optimal performance
- The environment variable JAVA_HOME should be set to the JDK installation direc
tory
   For example (linux): export JAVA_HOME=/usr/lib/jvm/java-6-sun
   For example (mac): export JAVA_HOME=/Library/Java/Home

Starting examples app...
2016-07-26 06:00:37,576 [main] INFO  DroolsExamplesApp started.
2016-07-26 06:00:37,657 [main] INFO  Found kmodule: jar:file:/Users/kkoitzsch/Do
wnloads/drools-distribution-6.4.0.Final/examples/binaries/drools-examples-6.4.0.
Final.jar!/META-INF/kmodule.xml
2016-07-26 06:00:37,871 [main] INFO  KieModule was added: ZipKieModule[releaseId
=org.drools:drools-examples:6.4.0.Final,file=/Users/kkoitzsch/Downloads/drools-d
istribution-6.4.0.Final/examples/binaries/drools-examples-6.4.0.Final.jar]
2016-07-26 06:00:37,871 [main] INFO  Found kmodule: jar:file:/Users/kkoitzsch/Do
wnloads/drools-distribution-6.4.0.Final/binaries/drools-pmml-6.4.0.Final.jar!/ME
TA-INF/kmodule.xml
2016-07-26 06:00:37,898 [main] INFO  KieModule was added: ZipKieModule[releaseId
=org.drools:drools-pmml:6.4.0.Final,file=/Users/kkoitzsch/Downloads/drools-distr
ibution-6.4.0.Final/binaries/drools-pmml-6.4.0.Final.jar]
▌
```

Figure A-8. *Successful installation and test of JBoss Drools*

Verifying the Environment Variables

Please insure the environment variable PROBDA_HOME, which is the root project directory, is set correctly in the .bash_profile file.

Basic environment variable settings are essential. Most of the components require basic variables to be set, such as JAVA_HOME, and the PATH variable should be updated to include the binary (bin) directories so programs can be executed directly. Listing A-1 contains a sample environment variable file as used by the example program. Other appropriate variables can be added as needed. A sample .bash_profile file is also provided with the online example code system.

Listing A-1. A sample of a complete environment variable .bash_profile file

```
export PROBDA_HOME=/Users/kkoitzsch/prodba-1.0
export MAVEN_HOME=/Users/kkoitzsch/Downloads/apache-maven-3.3.9
export ANT_HOME=/Users/kkoitzsch/Downloads/apache-ant-1.9.7
export KAFKA_HOME=/Users/kkoitzsch/Downloads/
export HADOOP_HOME=/Users/kkoitzsch/Downloads/hadoop-2.7.2
export HIVE_HOME=/Users/kkoitzsch/Downloads/apache-hive-2.1.0-bin
export CATALINA_HOME=/Users/kkoitzsch/Downloads/apache-tomcat-8.5.4
export SPARK_HOME=/Users/kkoitzsch/Downloads/spark-1.6.2
export PATH=$CATALINA_HOME/bin:$HIVE_HOME/bin:$HADOOP_HOME/bin:$ANT_HOME/bin:$MAVEN_HOME/
bin:$PATH
```

Make sure to run the Hadoop configuration script $HADOOP_HOME/libexec/Hadoop-config.sh when appropriate as shown in Figure A-9.

287

```
● ● ●                          ▓ hadoop-2.7.2 — -bash — 112×58
[Kerrys-MBP:hadoop-2.7.2 kkoitzsch$ ls                                                              ]
LICENSE.txt    bin          lib          sbin
NOTICE.txt     etc          libexec      share
README.txt     include      logs
[Kerrys-MBP:hadoop-2.7.2 kkoitzsch$ source $HADOOP_HOME/libexec/hadoop-config.sh                    ]
[Kerrys-MBP:hadoop-2.7.2 kkoitzsch$ printenv                                                        ]
HADOOP_DATANODE_OPTS=-Dhadoop.security.logger=ERROR,RFAS
SPARK_HOME=/Users/kkoitzsch/Downloads/spark-1.6.2
TERM_PROGRAM=Apple_Terminal
HADOOP_IDENT_STRING=kkoitzsch
SHELL=/bin/bash
TERM=xterm-256color
CATALINA_HOME=/Users/kkoitzsch/Downloads/apache-tomcat-8.5.4
HADOOP_HOME=/Users/kkoitzsch/Downloads/hadoop-2.7.2
TMPDIR=/var/folders/hz/nhnfch5j7vzdtw6vrxcrfkph0000gn/T/
HADOOP_PID_DIR=
Apple_PubSub_Socket_Render=/private/tmp/com.apple.launchd.ABCMnMZhrf/Render
HADOOP_PREFIX=/Users/kkoitzsch/Downloads/hadoop-2.7.2
TERM_PROGRAM_VERSION=361.1
OLDPWD=/Users/kkoitzsch
TERM_SESSION_ID=CE4DC6D4-172A-4758-B00A-8CF4C5F8C4D9
ANT_HOME=/Users/kkoitzsch/Downloads/apache-ant-1.9.7
USER=kkoitzsch
HBASE_HOME=/Users/kkoitzsch/Downloads/hbase-1.0.3
SSH_AUTH_SOCK=/private/tmp/com.apple.launchd.PCZAEM9Ki4/Listeners
MALLOC_ARENA_MAX=4
__CF_USER_TEXT_ENCODING=0x1F5:0x0:0x0
HADOOP_SECURE_DN_PID_DIR=
HADOOP_SECURE_DN_LOG_DIR=/
MAVEN_HOME=/Users/kkoitzsch/Downloads/apache-maven-3.3.9
PATH=/Users/kkoitzsch/anaconda/bin:/Users/kkoitzsch/Downloads/spark-1.6.2/bin:/Users/kkoitzsch/Downloads/apache-
tomcat-8.5.4/bin:/Users/kkoitzsch/Downloads/apache-hive-2.1.0-bin/bin:/Users/kkoitzsch/Downloads/hadoop-2.7.2/bi
n:/Users/kkoitzsch/Downloads/apache-ant-1.9.7/bin:/Users/kkoitzsch/Downloads/apache-maven-3.3.9/bin:/usr/local/b
in:/usr/bin:/bin:/usr/sbin:/sbin
HADOOP_HDFS_HOME=/Users/kkoitzsch/Downloads/hadoop-2.7.2
HADOOP_CLIENT_OPTS=-Xmx512m
HIVE_HOME=/Users/kkoitzsch/Downloads/apache-hive-2.1.0-bin
HADOOP_COMMON_HOME=/Users/kkoitzsch/Downloads/hadoop-2.7.2
PWD=/Users/kkoitzsch/Downloads/hadoop-2.7.2
HADOOP_YARN_HOME=/Users/kkoitzsch/Downloads/hadoop-2.7.2
JAVA_HOME=/Library/Java/JavaVirtualMachines/jdk1.8.0_102.jdk/Contents/Home
HADOOP_CLASSPATH=/Users/kkoitzsch/Downloads/hadoop-2.7.2/contrib/capacity-scheduler/*.jar
HADOOP_CONF_DIR=/Users/kkoitzsch/Downloads/hadoop-2.7.2/etc/hadoop
LANG=en_US.UTF-8
PROBDA_HOME=/Users/kkoitzsch/probda
XPC_FLAGS=0x0
HADOOP_PORTMAP_OPTS=-Xmx512m
HADOOP_OPTS= -Djava.net.preferIPv4Stack=true -Dhadoop.log.dir=/Users/kkoitzsch/Downloads/hadoop-2.7.2/logs -Dhad
oop.log.file=hadoop.log -Dhadoop.home.dir=/Users/kkoitzsch/Downloads/hadoop-2.7.2 -Dhadoop.id.str=kkoitzsch -Dha
doop.root.logger=INFO,console -Djava.library.path=/Users/kkoitzsch/Downloads/hadoop-2.7.2/lib/native -Dhadoop.po
licy.file=hadoop-policy.xml -Djava.net.preferIPv4Stack=true
HADOOP_SECONDARYNAMENODE_OPTS=-Dhadoop.security.logger=INFO,RFAS -Dhdfs.audit.logger=INFO,NullAppender
XPC_SERVICE_NAME=0
SHLVL=1
HOME=/Users/kkoitzsch
HADOOP_SECURE_DN_USER=
HADOOP_NAMENODE_OPTS=-Dhadoop.security.logger=INFO,RFAS -Dhdfs.audit.logger=INFO,NullAppender
HADOOP_MAPRED_HOME=/Users/kkoitzsch/Downloads/hadoop-2.7.2
```

Figure A-9. *Successful running of Hadoop configuration script and test with printenv*

Use "printenv" on the command line to verify default environment variable settings on start-up of a terminal window, as shown in Figure A-9.

References

Liu, Henry H. *Spring 4 for Developing Enterprise Applications: An End-to-End Approach.* PerfMath, http://www.perfmath.com. Apparently self-published, 2014.

Venner, David. *Pro Hadoop.* New York, NY: Apress Publishing, 2009.

■ ■ ■

Getting, Installing, and Running the Example Analytics System

The example system supplied with this book is a standard Maven project and may be used with a standard Java development IDE, such as Eclipse, IntelliJ, or NetBeans. All the required dependencies are included in the top-level pom.xml file. Download the compressed project from the URL indicated. Uncompress and import the project into your favorite IDE. Refer to the README file included with the example system for additional version and configuration information, as well as additional troubleshooting tips and up-to-date URL pointers. The current version information of many of the software components can be found in the VERSION text file accompanying the software.

Some standard infrastructure components such as databases, build tools (such as Maven itself, appropriate version of Java, and the like), and optional components (such as some of the computer vision–related "helper" libraries) must be installed first on a new system before successfully using the project. Components such as Hadoop, Spark, Flink, and ZooKeeper should run independently, and the environment variables for these must be set correctly (HADOOP_HOME, SPARK_HOME, etc.). Please refer to some of the references given below to install standard software components such as Hadoop.

In particular, check your environment variable PROBDA_HOME by doing a "printenv" command on the command line, or its equivalent.

For required environment variable settings and their default values, please refer to Appendix A.

Run the system by executing the Maven command on the command line after cd'ing to the source directory.

```
cd $PROBDA_HOME
mvn clean install -DskipTests
```

For additional configuration and setup information, see Appendix A.

For tests and example script pointers and directions, see the associated README file.

Troubleshooting FAQ and Questions Information

Troubleshooting and FAQ information can be referred to at the appropriate web page.

Questions may be sent to the appropriate e-mail address.

References to Assist in Setting Up Standard Components

Venner, David. *Pro Hadoop*. New York, NY: Apress Publishing, 2009.

© Kerry Koitzsch 2017
K. Koitzsch, *Pro Hadoop Data Analytics*, DOI 10.1007/978-1-4842-1910-2

Index

A

Algorithm
    coding, examples, 146
    survey, 139–141
    types, 139–141
Anaconda Python system
    initial installer diagram, 86
    installation, 87
Analytical engine
    rule control, 160
Analytic applications, 267, 268
Angular JS
    configuration file, 191–192
    console result, 194
    d3.js, 197
    directories and files, 187–188
    elasticUI, 186
    example system, 187
    graph database, 198
    handcraft user interfaces, 199
    JHipster, 186
    Maven stub, 190
    Neo4j, 199
    npm initialization, 188–189
    package.json file, 195
    sigma.js-based graph visualization, 198
    ./src/main/webapp/WEB-INF/
        beans.xml, 195–197
    ./src/test/javascript/karma.conf.js, 193
ANSI SQL interface and
        multi-dimensional analysis, 54
Apache Beam, 80, 268
Apache Bigtop, 61–62
Apache Calcite, 74–75
Apache Cassandra, 73
Apache Falcon, 82
Apache Flink, 80
Apache Hadoop, 80, 268
Apache Hadoop Setup, 280
Apache Kafka, 12, 80
Apache Kafka messaging system, 60

Apache Katta
    configuration, 96–97
    initialization, 96–97
    installation, 96–97
    solr-based distributed data pipelining
        architecture, 96
Apache Kylin, 74
Apache Lens (lens.apache.org)
    Apache Zeppelin, 72
    architecture diagram, 70–71
    installed successfully using
        Maven on MacOSX, 71
    login page, 72
    OLAP commands, 71, 74
    REPL, 71–72
    zipped TAR file, 71
Apache Lenya, 268
Apache Lucene, 16
Apache Mahout
    classification algorithms, 54
    and Hadoop-based machine
        learning packages, 54
    software frameworks, 54
    in visualization, 55
    Vowpal Wabbit, 54
Apache Maven, 44–45
Apache MRUnit, 61
Apache NiFi, 268
Apache Phoenix, 18
Apache POI, 182
Apache software components, 267
Apache Solr, 16
Apache Spark, 8, 10–11, 13, 18, 22–26, 80
Apache Spark applications, 73
Apache Spark-centric technology stack, 142
Apache Spark libraries
        and components
    different shells to choose from, 56
    Sparkling Water (h20.ai), 58
    H20 Machine Learning, 58
    Python interactive shell, 56
    streaming, 57

© Kerry Koitzsch 2017
K. Koitzsch, *Pro Hadoop Data Analytics*, DOI 10.1007/978-1-4842-1910-2

Get the eBook for only $4.99!

Why limit yourself?

Now you can take the weightless companion with you wherever you go and access your content on your PC, phone, tablet, or reader.

Since you've purchased this print book, we are happy to offer you the eBook for just $4.99.

Convenient and fully searchable, the PDF version enables you to easily find and copy code—or perform examples by quickly toggling between instructions and applications.

To learn more, go to http://www.apress.com/us/shop/companion or contact support@apress.com.

Printed in the United States
By Bookmasters